意大利的四季滋味

[意] 马里诺·安托尼奥（Marino D'Antonio） 著

张际星 译

电子工业出版社

Publishing House of Electronics Industry

北京·BEIJING

自序 Preface

身为厨师，最令我感到满足的事情就是，有幸能通过烹饪来感受大自然给予人们的美好馈赠，同时，借助自己的双手和想象力，使美好的食材与健康的美食结缘，将爱与专注融汇其中，为我们的家人、朋友或客人献上各种珍馐佳肴。在这本书里，您可以学到很多有趣的意大利饮食文化和家常食谱，了解美食背后那些有趣的故事。

食物有着一种特殊的魔力，能够将不同民族、不同地域、不同文化背景的人们聚集在一起，一同分享美味、趣闻和欢笑，生命中许多重要的和美好的时刻与回忆，常常就发生在餐桌上。

烹饪是交流，是记忆，是文化，又是历史。

我希望通过书中的文字与图片，您能够深入了解意大利传统菜肴、意大利饮食与文化，并通过饮食的脉络去了解意大利悠久的历史。我也希望把本书当作一个媒介，帮助人们分享美食，并创造美好的心境。

从传统意义上说，意大利饮食的关键元素是谷物、蔬菜、橄榄油和肉类。其中，面包、意大利面、番茄、蔬菜、橄榄油和奶酪是意大利菜的基本成分。同时，丰富而新鲜的香草，如迷迭香、罗勒、欧芹等，对意大利菜的烹调也起到了重要作用。

如果厨师对食材和传统食谱具有深刻的理解，并掌握了熟练的"厨艺语言"，就能制作出世界级的意大利风味美食。同时，厨师还可以假以当代最新时尚和创新手法来二次创新或重构一些百年经典食谱，利用现代烹饪技法使菜肴更清淡，更适合精致的味觉，这其中最重要的还是创新和创意，它们与传统可以完美地融汇相承。

美食是意大利饮食文化的核心，意大利人非常重视他们的饮食文化，一个典型的表现便是他们对食谱和烹饪技艺的痴迷。很多时候，你会看见意大利人热烈地讨论某道菜的起源以及正确烹饪方式，对食物的依恋早已注入到了意大利人的血液之中。未来，我的努力方向主要是：首先设法提升我的料理品质，同时尽可能地减少碳排放，减少对进口食材的依赖，增加本地食材的使用；其次还要注意保护环境，实现美食及其文化的可持续发展，通过自己的行动，保护环境、保护地球。

我很幸运，在过去的十五年里，见证了中国日新月异的发展历程。中国地大物博，我非常喜欢在料理中运用中国的本地食材，比如云南的香草和蔬菜，中国有些产地的鱼子酱品质比欧洲的更优。让我来举出一些实例：酱油、云南香菇、山东韭菜、黑龙江鱼子酱、新疆葡萄干和无花果、大连扇贝和海胆、厦门鱿鱼和小章鱼、山西醋、贵州黑猪肉、四川辣椒……这些美好的食材总能激励我创新和制作新的菜品，它们也能让您和我对未来怀有更多的期待。

Marino 于北京

2020 年 7 月

One of the most fulfilling things in a chef's life is the opportunity to work with the gifts that Mother Nature gives us. It is a joy to use our hands and our imagination to combine beautiful ingredients and prepare delicious dishes for our friends and family. In this book, you will discover some fun Italian family recipes, and the stories behind them.

Food has the power to bring people from different nations, regions and cultural backgrounds together. While we are enjoying a great meal, we can share stories and laughter. The dinner table sees many great moments and memories in our lives.

Cooking is communication, memory, culture, and history.

Through these recipes, I hope you will gain a better vision of traditional Italian dishes and culture, and will learn something about Italy's profound history. I hope this book will be a medium for people to share good food and good times together.

Traditionally, the key components of Italian cuisine are cereals, vegetables, olive oil, bread, pasta, tomatoes, and cheese are also fundamental ingredients. Fresh herbs such as rosemary, basil, and parsley also play an important role.

Italian cuisine can be executed at a world-class level when a chef has extensive knowledge of ingredients and traditional recipes. Meanwhile, with modern and creative techniques, chef can reinvite centuries-old recipes and make dishes taste lighter and more appealing to refined palates. Nevertheless, what matters most is retaining the spirit of the original recipe with creative ideas.

Food is the core of Italian culture. Italians are very attached to their regional recipes. Taking a typical example, they often get into heated arguments about the origin of a recipe and how to execute it in the most proper way. This respect and love for food is in our blood. I constantly try to improve the quality of my cooking and reduce my carbon footprint by switching out some imported ingredients for local ones. I pay attention to the environment and pursue the sustainable development and culture of Italian cuisine and protect the enviornment and our earth through my actions.

I feel so lucky that I have been able to witness some fascinating changes here in China over the past 15 years. With its vast territory and abundant resourses, China is now producing excellent local ingredients. For example, there is high-quality herbs and vegetables in Yunnan, and the local caviar is now better than European brands. To give a few more concrete examples, ingredients like soy sauce and mushrooms from Yunnan, leeks from Shandong, caviar from Heilongjiang, raisins and figs from Xinjiang, scallops and sea urchins from Dalian, squid and baby octopus from Xiamen, Shanxi vinegar, black pork from Guizhou, and even chilies from Sichuan... inspire me to do new research and create more dishes, I expect we will have many more ingredients to look forward to in the future.

Marino, Beijing

July, 2022

推荐序

大家熟悉的"老马"意大利名字叫 Marino D'Antonio，他来中国已经十几年了。他是一个热爱意大利也热爱中国的"老外"，他长居中国，有许多中国朋友和众多的中国粉丝。他在北京侨福芳草地的 Opera BOMBANA 餐厅任行政总厨有很长时间，赢得了很多赞誉和人缘。

老马很活跃，能折腾，从他不断地换工作、换城市、开餐厅就能看得出来。老马每次开新餐厅，我都会第一时间到场尝鲜，老马是我的好朋友，不到说不过去。

两年前，老马回到北京，在三里屯北街的瑜舍酒店一层，主理一家意大利的"顺口菜"餐厅——FRASCA。我记得当时吃了一道炸小海鲜：炸小鱼小虾，酥酥的，配着蛋黄酱、酸黄瓜和番茄酱，很好吃。我们还吃了一款龙虾面，特有新意的是，上面撒了干鱼子碎，也有说是海胆碎的，反正都好吃。烤 T 骨牛排很有肉质感。我甚至觉得，吃这种有劲的牛肉，应该能变成大力水手。

2021 年 7 月的一天，老马在中国大饭店三层的 GIADA Garden 迦达花园餐厅担任行政总厨。当晚，我吃了好几道菜都非常好——低温的南极虾质柔润色如玉，深海鳕鱼口感滑润，手工管面依然功底扎实，低温牛小排酥烂。这是我近几年吃到的最好的西餐，这家餐厅也成了我必须来的地方。

印象最深的是老马给我做了他家乡最独特的 Pinsa Romano（音"拼萨"），面皮酥脆是其最明显的特点，马苏里拉芝士拉出很长的丝。

我们吃了一款芝士、罗勒和樱桃番茄的，还有一款帕尔马火腿卷芝麻菜的。我问老马 Pinsa 的特点，尤其提到酥酥的面皮，老马说："Pizza 和 Pinsa 的区别在于，Pizza 用的是高筋面粉，含水量超过 70%，发酵时间至少为 48 小时；而 Pinsa 面粉是由不同种类的谷物混合而成的，如大米、大豆和斯佩耳特小麦面粉，且发酵时间更长。Pizza 是意大利南部那不勒斯地区的美食，而 Pinsa 来自意大利中部罗马地区。"

每次来老马的餐厅吃饭，他都忙进忙出的，介绍完菜品，就会跑回厨房准备下一道菜，最后还会特郑重地让我提提意见。我说："每一次都吃得很舒服。"老马兴奋地说："对对，我就是要做让北京人感觉特舒服的正宗意大利菜。"我说："Pinsa 酥酥的口感，我很喜欢。"他睁大眼睛说："真的吗？"我也郑重地说："真的，我想吃 Pinsa 的时候，就不用去意大利了。"

我还对老马说，听说你家里人都会做饭，老马听了特自豪。我接着说："最后还有一句话，听说在你们家中，你的厨艺是最差的！"老马听了差点晕了。哈哈。

老马写的美食书，和他做的菜一样，舒舒服服，没有压力，没有距离，能在字里行间感受到他的细腻和热情，读着，读着，就想家了……

大董

2021 年 10 月

推荐序

Marino 要出第二本书了，新书按照四季不同时令设计出当季食谱，和中国传统的"应时应季，不时不食"理念相符合，并根据西方的饮食习惯以套餐菜单的形式呈现。

我在平时的教学中尤其是拓展模块方面，就是以菜单形式对学生进行实践训练的，主要特点就是尊重文化、符合规律、依照习惯、综合训练、模拟仿真，并且效果良好。

《意大利的四季滋味》一书形式新、体例新、菜品美观、传统创新兼顾，体现了 Marino 这么多年对职业的敬畏和技术经验的积累。

和 Marino 认识十多年了，他是我接触的外籍厨师中和我关系比较密切的，用北京话说就是"铁哥们儿"，一方面因为他是著名米其林大厨 Bombana 先生在北京餐厅开业时的首任行政主厨，那个时候北京侨福芳草地是时尚的美食热点坐标。另一方面，和他一起工作的厨师很多都是我的学生，无论是在 Sureno、Opera BOMBANA、北京瑜舍酒店 FRASCA 意大利餐厅，还是在中国大饭店三层的 GIADA Garden 迦达花园餐厅。

在 Opera BOMBANA 餐厅，Marino 任行政总厨，我几乎参加了该餐厅的全部年度店庆活动、每轮菜单的更换品鉴、"大神"Bombana 先生每次到访，以及有关餐厅运营的各种推广活动，包括每年的法国及澳洲冬季松露季、白松露季，与江振诚、Riccardo、BINBIN 等米其林主厨联袂合作等活动……在和 Marino 的沟通和交流中，我们逐渐成了非常要好的朋友。

我在欧洲学习时，最开始就是在意大利餐厅工作。意大利是一个爱好美食的国度，饮食方面有着悠久的历史，从罗马帝国时期、中世纪时期到文艺复兴时期，一直都是欧洲繁荣、富庶的地区。在这片土地上，诞生了米开朗基罗、拉斐尔、达·芬奇、薄伽丘、马萨乔、乔托等艺术、绘画、雕塑、文学、哲学和建筑等众多门类的大师与巨匠，同时孕育了丰富的意大利饮食文化。

意大利各地的饮食，也呈现出不同的特色，如同他们的古今艺术、时装、足球和现代汽车，闪耀着光彩。位于亚平宁半岛的罗马、米兰、拿波里、西西里、托斯卡纳、佛罗伦萨、威尼斯、博洛尼亚、热那亚等地物产丰富，勤劳、勇敢、智慧的意大利人民奉献出意大利面、橄榄油、海鲜、奶酪、番茄、比萨饼、帕尔马火腿、白松露、葡萄酒、冰激凌、提拉米苏……这些我们耳熟能详的美食。意大利美食既典雅高贵，又厚重真朴，讲究原汁原味，在世界各地的影响力极大，受到世界各地人们的喜爱。

应该是在北京工作的原因，Marino 为人非常有"老北京范儿"——见多识广，局气厚道，有礼有"面儿"，宽容和善。他是在北京我认识的外籍厨师中，非常注重吸收当地文化，学习本地风俗，听取他人意见，并且能够学以致用、扬长避短，融入中国生活节奏，和顺友善的意大利人，在很多方面我们有相同的价值观。感谢 Marino 的新书，一如其人，专业、讲究、深刻、有趣，相信它对专业厨师、相关专业的学生和美食爱好者都有所帮助，Grazie Mille！

侯德成

2022 年 5 月

推荐序

　　Marino 音译成中文的名字是马里诺，记得有一次我和他一起去"大董"餐厅参加晚宴，大董老师说马里诺叫着拗口，不如直接叫"老马"来得亲切，于是"老马"这个名字就在中国的餐饮圈儿叫开了。越仔细想，越觉得"老马"确实太符合 Marino 的气质了：亲切、热情、实在、重情谊、讲义气、接地气……

　　老马从来没有把自己当成"老外"，也许正因为这一点，老马的中国哥们儿特别多，我和老马认识快十年了，他是我的良师益友和至亲家人。

　　我和老马相识于 Opera BOMBANA 餐厅。2013 年，这家餐厅开业，这也是北京第一家真正拥有米其林三星"血统"的意大利餐厅，也是从那时起，我开始全面地感悟米其林餐厅的台前幕后和意大利美食文化，可以说老马是我的启蒙老师。我将所看到的和所学到的鲜为人知的餐厅有趣故事和意大利饮食知识综合起来，筛选后再转述给媒体和大众，这就是我在餐厅开业初期时的主要工作。

　　我记得，一位媒体朋友想写一篇关于意大利面的文章，网上的资料有限，她便跑到 Opera BOMBANA 餐厅来求助，我和老马发现她从网上找的资料很多都有错误，且大多都是笼统的泛泛之谈，这促使我们萌生了撰写一本意大利美食书的念头，于是老马按照"美食地图"的思路，创作了他的第一本书——《来吃意大利菜》。在书中，他将意大利分成东北、西北、中部、南部四大区域，同时还将意大利与中国相似区域的饮食做了对比，比如意大利东北地区偏冷，人们爱吃山货（菌菇）、意大利"棒子面"（Polenta）、羊肉炖菜、牛羊肉馅饺子等，这些像极了中国的东北菜，书中有很多诸如此类有趣的故事。

老马很重情谊，他在 Opera BOMBANA 餐厅一干就是七年。2018 年他离开的时候没有带走一名员工，为了不带走老客人，他竟然只身前往上海工作。三年后，老马重归京城，一手打造 GIADA Garden 迦达花园意大利餐厅，与此同时，我们合作的第二部美食图书项目也重新启动。

　　回到北京后的老马更加沉稳成熟，更加珍惜老朋友，他对第二本书有了更多新的想法，这本书不仅讲述了意大利的岁时四季，更是通过时令与食物的关系，记叙了意大利的人文情谊，书中呈现了一次次节假日聚餐的场景，承载着他对家人和朋友满满的爱和思念。希望大家读完这本书再去意大利的时候，在欣赏那里的绝美建筑和自然风光之余，更要慢下来，用心品尝意大利的当季美食，体验意大利人的生活哲学。

张际星

2022 年 7 月

目录

马里诺·安托尼奥（Marino D'Antonio），意大利名厨，现任 GIADA Garden 迦达花园餐厅行政总厨。来自意大利北部城市贝加莫，师从米其林三星名厨，在中国生活十余年，2016 年出版第一部作品《来吃意大利菜———一场华丽的美食行走》。在高端餐饮行业拥有超过 20 年工作经验，曾任北京 Opera BOMBANA 意大利餐厅行政主厨、上海镛舍酒店及北京瑜舍酒店 FRASCA 意大利餐厅行政总厨，荣登"2020 上海米其林指南"榜单，屡获 *Time Out Beijing*，*The Beijinger* 等杂志"年度主厨 Chef of the Year"的殊荣。

老马和 GIADA Garden 迦达花园餐厅的厨师团队

老马和甜品师 Filippo Mazzanti

Spring
春

三月

Marzo
March

三月，对意大利人来说并不是那么令人兴奋的时节，冬日的寒冷刚刚退却，乍暖还寒。不管怎样，在三月的意大利人的餐桌上，还是能看到些许春天的颜色：青绿的罗勒叶、五色的彩椒、金黄色的鸡汤和鸡蛋薄饼……虽不如夏天那般绚丽灿烂，但还是令人感受到了生机萌发的喜悦。另外，吃一份提拉米苏，味蕾的美好带来勇气和热量，用来抵御依然寒风凛冽的三月。

March is not an exciting month in Italy. The winter chill is just over, but the warm weather is still a few months away. Nevertheless, spring colors can be found on the dining table: fresh green basil, rainbow-coloured peppers, golden chicken stock and egg pancakes... Although March is not as brilliant as summer, it is still full of the joy of germination. This month, it is nice to enjoy a rich tiramisu that will give you the strength to get through the windy days.

Menu

新西兰鳌虾配茴香头及香橙沙拉
Finocchi Scampi e Arance
Scampi, Fennel Bulb and Orange Salad

有机鸡汤配鸡蛋薄饼
Crespelle in Brodo
Paper-Thin Pancakes in Chicken Consommé

意式土豆球配番茄及新鲜水牛芝士
Gnocchi alla Sorrentina
Potato Gnocchi with Tomato and Buffalo Mozzarella

智利黑鳕鱼配彩椒及橄榄
Merluzzo Peperoni Olive
Cod Fillet with Bell Peppers and Olives

提拉米苏
Tiramisu
Tiramisu

FINOCCHI SCAMPI E ARANCE

SCAMPI, FENNEL BULB AND

ORANGE SALAD

新西兰鳌虾配茴香头及香橙沙拉

茴香头 3 个	3 fennel bulbs
柠檬 1 片	1 lemon slice
新西兰鳌虾 12 只	12 New Zealand scampi
橙子 2 个	2 oranges
番茄 3 个	3 tomatoes
罗勒 20 克	20g basil
柠檬汁 2 勺	2 tsp lemon juice
橄榄油 100 克	100g olive oil
橙子酱汁 60 克	60g orange dressing
奥西特拉鱼子酱 40 克	40g Oscietra caviar
盐适量	Salt, to taste
白胡椒适量	White pepper, to taste
罗勒叶，装饰用	Basil leaves, to garnish

制　　作　　洗净茴香头，并用刨片器将其刨成薄片。

准备一碗冰水，放上一片柠檬，把茴香头片放在里面浸泡。

鳌虾洗净后上锅蒸 1 分钟，如果你喜欢的话，可以保留虾尾。待虾冷却，再用橄榄油、盐、白胡椒和少许柠檬汁调味。

将橙子去皮去核，将橙子果肉按瓣分好，去掉橘络和白色的膜。

将番茄在热水中烫一下之后去皮，再用勺子把中间的籽去除，切成丁，用盐、切碎的罗勒和橄榄油调味。

将茴香头片切丝。在沙拉碗中将茴香头丝、橄榄油、柠檬汁、盐和白胡椒混合。

把调味好的茴香头沙拉、橙子块和鳌虾摆放在盘子内，用鱼子酱点缀其上，并装饰上一些橙子块，再用一些罗勒叶、番茄丁装饰。

Prepare
the dish　　Wash the fennel bulbs. Thinly slice them on a mandolin.

Place the fennel bulbs in a bowl of iced water with the slice of lemon.

Clean the scampi and steam for 1 minute, keeping the tail on, if you like. Let the scampi cool and then season with olive oil, salt, white pepper, and some lemon juice.

Peel and core the orange, then separate the flesh into segments, removing the white pith and membrane.

Blanch the tomato to remove the skin and then scoop out the seeds. Cut the flesh into cubes and season with salt, chopped basil, and olive oil.

Cut the fennel bulbs into shreds. Put the fennel in a mixing bowl and add olive oil, lemon juice, salt, and white pepper.

Arrange the fennel bulbs salad, some orange supremes, and the scampi neatly on the plate.

Finish the dish with caviar, a few drops of orange dressing, the diced tomato, and garnish with some nice basil leaves.

CRESPELLE IN BRODO
PAPER-THIN PANCAKES IN
CHICKEN CONSOMMÉ

有机鸡汤配鸡蛋薄饼

薄饼 | **For the pancakes**
面粉 30 克 | 30g flour
鸡蛋 3 个 | 3 eggs
水 100 毫升 | 100ml water
牛奶 100 毫升 | 100ml milk
帕玛森芝士 80 克 | 80g Parmigiano Reggiano
盐适量 | Salt, to taste
黑胡椒适量 | Black pepper, to taste
肉豆蔻粉适量 | Ground nutmeg, to taste
橄榄油适量 | Olive oil, to taste

鸡汤 | **For the consommé**
芹菜 150 克 | 150g celery
胡萝卜 150 克 | 150g carrots
洋葱 150 克 | 150g onions
鸡胸肉 300 克 | 300g chicken breast
蛋白 200 克 | 200g egg whites
鸡汤 1 升 | 1L chicken stock

薄　　饼	将面粉、鸡蛋、盐、黑胡椒和肉豆蔻粉放在盆中，加入水和牛奶，手动混合搅拌均匀。用细筛过滤面糊后静置 30 分钟。
	用中火加热不粘锅（在烹制过程中也尽量保持该温度）。
	用厨房纸蘸水擦一下锅，再倒上少许橄榄油。倒上一些面糊，尽可能地摊成很薄的饼。
	将饼铺在砧板上，撒上帕玛森芝士和肉豆蔻粉，像雪茄一样卷起来。

Prepare
the
pancakes

Place the flour, eggs, salt, black pepper and ground nutmeg in a mixing bowl. Add the water and the milk, and whisk well to combine. Pass through a very fine sieve and rest for 30 minutes.

Heat a nonstick pan over a moderate heat (try to keep the temperature even during cooking).

Wet the pan with some paper tissue and add a couple of drops of olive oil. Add some batter, and make the pancake as thin as you can.

Spread each cooked pancake on a board and sprinkle with Parmigiano Reggiano and ground nutmeg, then roll up like a cigar.

鸡　　汤

将所有蔬菜洗净并切好，把它们同鸡胸肉和蛋白一同倒入食品料理机中，搅拌至质地细腻均匀。

在混合物中加入鸡汤，文火炖煮 1 小时。

当鸡汤变得清澈透明，用大号咖啡滤纸进行滴滤过滤。

重新加热鸡汤，并确认下味道。在盘中放入 3 张薄饼，倒入热鸡汤。

Prepare
the
consommé

Clean and peel all the vegetables and chop them in a food processor with the chicken and the egg white until you get a soft, homogeneous paste. Put the mixture in a pot with the cold chicken stock and bring to a simmer over a low heat, cooking gently for 1 hour.

When the stock is crystal clear, pass it through a large coffee filter and leave it to drip.

Reheat the consommé and check the seasoning. Serve three pancake rolls per plate and cover with hot chicken consommé.

GNOCCHI ALLA SORRENTINA
POTATO GNOCCHI WITH TOMATO AND BUFFALO MOZZARELLA
意式土豆球配番茄及新鲜水牛芝士

土豆球	For the gnocchi
土豆 1 公斤	1kg potatoes
帕玛森芝士 200 克	200g Parmigiano Reggiano
面粉 300 克	300g flour
蛋黄 2 个	2 egg yolks
海盐适量	Sea salt, to taste
黑胡椒适量	Black pepper, to taste

酱汁	For the sauce
大蒜 2 瓣	2 garlic cloves
樱桃番茄 400 克	400g cherry tomatoes
番茄酱 150 克	150g tomato sauce
水牛芝士 200 克	200g buffalo mozzarella
帕玛森芝士 120 克	120g Parmigiano Reggiano
新鲜罗勒 30 克	30g fresh basil
橄榄油适量	Olive oil, to taste
盐适量	Salt, to taste
黑胡椒适量	Black pepper, to taste

土 豆 球	烤箱预热至 200℃。将土豆放入烤盘内，加少许水，撒上少许海盐，烤至土豆变软。

趁土豆还热的时候，将其去皮、碾成泥。加入帕玛森芝士和面粉，再加入蛋黄、海盐和黑胡椒。用手将所有食材混合，直至揉成光滑的面团，然后再撒一些面粉。

在面团仍旧温热时，将其全部切成 1 厘米厚的小面团。随后在砧板上撒一些面粉或粗麦粉，将切好的小面团放在砧板上备用。

Prepare the gnocchi

Preheat the oven to 200℃. Put the potatoes in an oven tray with some water and top each potato with a pinch of sea salt. Cook until soft.

While the potatoes are still warm, remove the skin and pass them through a potato ricer. Add the Parmigiano Reggiano and the flour, then the egg yolks, sea salt, and black pepper. Mix by hand until the mixture comes together into a smooth dough and then sprinkle with some flour.

While the dough is still warm, cut it into 1cm cubes. Leave the gnocchi to rest on a wooden surface sprinkled with flour or semolina.

酱 汁

樱桃番茄洗净，每个切成 4 瓣。水牛芝士切成小丁。

在煎锅中放入橄榄油和大蒜，将大蒜煸炒至金黄色，用漏勺将大蒜捞出。放入樱桃番茄，翻炒 5 分钟，加入番茄酱，用盐和黑胡椒调味。

将一锅水煮沸，放入土豆球，当其浮在水面上的时候就说明煮好了，捞出后用制作好的酱汁拌匀。将拌好的土豆球放入烤盘，上面铺上水牛芝士和帕玛森芝士，放入烤箱 200℃烤制 3 分钟。取出后点缀适量新鲜罗勒。

Prepare the sauce

Wash the cherry tomatoes and cut each one into quarters. Cut the buffalo mozzarella into small dice.

Heat the olive oil and garlic in a sauté pan and cook the garlic until golden brown, then remove with a slotted spoon. Add the cherry tomatoes and cook for 5 minutes. Add the tomato sauce and season with salt and black pepper.

Bring a large pot of water to the boil and add the gnocchi. The gnocchi are ready when they float to the surface. Strain the gnocchi and add to the tomato sauce.

Transfer the gnocchi and tomato sauce to a baking dish, top with the buffalo mozzarella cubes and the Parmigiano Reggiano, and bake at 200℃ for 3 minutes. Serve with some fresh basil tips on top.

MERLUZZO PEPERONI OLIVE
COD FILLET WITH BELL PEPPERS
AND OLIVES

智利黑鳕鱼配彩椒及橄榄

酱汁	For the sauce	鳕鱼柳	For the cod fillets
橄榄油 50 克	50g olive oil	鳕鱼柳 4 块（200 克 / 块）	4 cod fillets (200g each)
洋葱 50 克，切碎	50g onions, chopped	黑醋 80 克	80g balsamic vinegar
辣椒 1 个	1 chilli	橄榄油 40 克	40g olive oil
鱿鱼圈 100 克	100g calamari rings	盐适量	Salt, to taste
橄榄 100 克，去核	100g olives, pitted	黑胡椒适量	Black pepper, to taste
彩椒汁 250 毫升	250ml bell pepper juice	应季蔬菜	Seasonal vegetables
盐适量	Salt, to taste		
黑胡椒适量	Black pepper, to taste		

酱　　汁　　中火热锅，加入橄榄油和洋葱碎，略炒几分钟后加入辣椒、鱿鱼圈和橄榄。
烹制 10 分钟后，加入彩椒汁，再煮 10 分钟，搅拌酱汁直至润滑。用细筛过滤后，用盐、黑胡椒调味，之后保温放置。

Prepare the sauce

Heat a large pot over a medium heat, then add the olive oil and the onions. Cook the onions for a few minutes then add the chilli, the calamari rings, and the olives.
Cook for about 10 minutes and then add the bell pepper juice. Simmer for another 10 minutes and then blend the sauce until smooth. Pass through a fine sieve. Season with salt and black pepper, and keep warm.

鳕 鱼 柳　　用黑醋、盐和黑胡椒腌制鳕鱼柳。
在铸铁锅内加入橄榄油，放入鳕鱼柳略煎。将鳕鱼柳放入预热好的烤箱，200℃烤 3 分钟。
将酱汁点缀在盘子中，将鳕鱼柳放在盘中间，用应季蔬菜做装饰即可。

Prepare the cod fillets

Marinate the cod fillets in the balsamic vinegar, salt, and black pepper.
Heat some olive oil in a cast iron pan and roast the fish for a few minutes. Finish in a 200°C oven for 3 minutes.
Sauce each plate and place the roasted fish in the middle, and then garnish with seasonal vegetables.

TIRAMISU

TIRAMISU

提 拉 米 苏

萨瓦多海绵饼干
糖粉 30 克
蛋糕粉 35 克
马铃薯淀粉 35 克
蛋白 80 克
白砂糖 30 克
蛋黄 90 克

马斯卡彭奶油
巴氏杀菌蛋黄 150 克
白砂糖 100 克
干玛萨拉酒 60 克
马斯卡彭奶酪 500 克

咖啡糖浆
新鲜意式浓缩咖啡 100 克
白砂糖 20 克
可可利口酒 15 克
杏仁力娇酒 20 克

马斯卡彭冰激凌
白砂糖 150 克
葡萄糖 220 克
牛奶 960 克
干玛萨拉酒 70 克
转化糖浆 40 克
蛋黄 110 克
马斯卡彭奶酪 360 克
可可粉适量

For the savoiardi biscuits
30g icing sugar
35g cake flour
35g potato starch
80g egg whites
30g sugar
90g egg yolks

For the mascarpone cream
150g pasteurised egg yolks
100g sugar
60g dry Marsala wine
500g mascarpone

For the coffee syrup
100g fresh espresso coffee
20g sugar
15g Kahlua
20g Disaronno

For the mascarpone gelato
150g sugar
220g dextrose
960g milk
70g dry Marsala wine
40g Trimoline
110g egg yolks
360g mascarpone
Cocoa powder, to taste

萨 瓦 多　　在碗里把糖粉、蛋糕粉和马铃薯淀粉混合在一起。
海绵饼干　　将蛋白和白砂糖打发至稳固，然后拌入蛋黄，混合均匀后，分两次加入面粉混合物，搅拌均匀。
　　　　　　将混合物放入裱花袋中，在烤盘中挤出长条饼干形状。在饼干表面撒上糖粉，烤箱预热至 160℃，
　　　　　　烘烤饼干约 10 分钟，直到饼干变成浅金色。
　　　　　　待饼干冷却，保存在密封容器中，在室温下放置。

Prepare　　In a bowl, mix the icing sugar, cake flour, and potato starch together.
the　　　　In the bowl of a stand mixer, whip the egg whites with the sugar to form stiff peaks then fold in the
savoiardi　　egg yolks. Fold the dry mixture into the egg mixture in two batches.
biscuits　　Transfer the mixture to a piping bag and pipe some long strips. Dust the strips with icing sugar
　　　　　　before baking them at 160℃ in a static oven until they are lightly golden, around 10 minutes.
　　　　　　Cool the sponge biscuits and reserve in an airtight container at room temperature until you are
　　　　　　ready to use them.

马斯卡彭　　用搅拌机把蛋黄和白砂糖打发至浓稠，加入干玛萨拉酒，继续搅拌。再加入马斯卡彭奶酪，继续
奶　　油　　搅拌，直到整个混合物混合均匀。把奶油放在冰箱里冷藏。

Prepare the　In the bowl of a stand mixer, whip the egg yolks with the sugar to a ribbon stage. Add the dry
mascarpone　Marsala wine and continue mixing. Once combined, add the mascarpone in small batches and
cream　　　whip for few minutes until the whole mixture is firm. Reserve the cream in the fridge until ready to
　　　　　　use.

组　　装　　在蛋糕模具底部铺上一层萨瓦多海绵饼干，刷上混合好的咖啡糖浆，然后涂上适量马斯卡彭奶油。
　　　　　　再放一层饼干，刷上咖啡糖浆，再抹上马斯卡彭奶油。 把提拉米苏放在冰箱里冷藏，直到准备装
　　　　　　盘。

Assembly　　Place a layer of savoiardi biscuits in the bottom of a cake tin, using a brush to wet them with
　　　　　　coffee syrup. Spread some mascarpone cream over. Repeat and finish with a layer of mascarpone
　　　　　　cream. Keep the tiramisu in the fridge until ready to serve.

马斯卡彭
冰激凌

在大碗里混合白砂糖和葡萄糖。

锅中倒入牛奶、干玛萨拉酒和转化糖浆，开始加热。当混合物温度在 40℃时，加入蛋黄继续搅拌，之后慢慢倒入白砂糖和葡萄糖的混合物，继续搅拌。用温度计测量温度，一旦混合物达到 82℃即可停止加热，并在容器中保存至少一个晚上。

第二天加入马斯卡彭奶酪，然后在搅拌机中搅拌冰激凌，之后放在冰箱里备用。

Prepare the
mascarpone
gelato

In a large bowl, combine the sugar and dextrose and mix well.

Pour the milk, dry Marsala wine and Trimoline into a pot and start to heat it. When the mixture starts to get warm at around 40℃, add the egg yolks and keep mixing. Then, start to slowly add the dry ingredients. Continue mixing and check the temperature with a digital thermometer. Once the mixture reaches 82℃, take it off the heat and reserve it in a container for at least one night. The next day, add the mascarpone before proceeding to churn the gelato. Transfer to the freezer and keep until ready to use.

摆　　盘

用一只大勺子舀一大块提拉米苏在盘子里，旁边放一层马斯卡彭冰激凌，最后撒上可可粉。

Plating

Using a large spoon, scoop a big portion of tiramisu onto each plate. Place a quenelle of the mascarpone gelato next to it and dust everything with cocoa powder.

四月

Aprile
April

四月的意大利，大自然正在苏醒，你可以看到崭新的色彩，各色时蔬也在肆意生长。白芦笋是当下最美味的食材之一，有很多种烹饪方法，无论怎么做，都从来不会令你失望。此刻，花园里已经发现了今年生长的第一批节瓜，这是意大利人非常喜欢的一种蔬菜，可以烹饪很多种美味，还可以储存过冬。

In April, nature is waking up and you can see new colours and beautiful vegetables growing all around you. White asparagus is one of the most delicious ingredients available this month. This elegant vegetable can be prepared in many different ways and never disappoints. In addition, the first zucchini can be found growing in the garden, and many people start preserving them for the winter.

同样在四月，意大利人会同家人和朋友聚餐庆祝复活节，这可是寒冷季节过后享受户外温暖阳光和空气的好时机。传统意义的复活节星期一被称为"Pasquetta"，户外野餐是必须的，复活节巧克力蛋当然也是不可缺少的传统，这可是孩子们和巧克力爱好者的盛宴！

Italians also celebrate Easter in April, the perfect occasion to gather together with family and friends to eat and enjoy the open air after the cold winter. Traditionally, Easter Monday is called "Pasquetta", and picnics are a must. Of course, it wouldn't be Easter without chocolate eggs. Kids and chocolate lovers alike delight in Easter egg hunts.

Menu

意大利煎蛋饼
Frittata di Pasta
Italian Omelette with Spaghetti, Zucchini and Bacon

白芦笋汤配香煎北海道扇贝
Asparagi Bianchi e Capesante
White Asparagus Soup with Seared Scallops

意式扁面配帝王蟹和节瓜
Linguine con Granchio e zucchinine
Linguine with King Crab and Zucchini

多宝鱼柳配茴香及西西里风味茄子酱
Rombo Finocchi e Caponata Siciliana
Turbot Fillet with Fennel and Sicilian-Style Caponata

巧克力复活节彩蛋
Chocolate Easter Eggs

FRITTATA DI PASTA
ITALIAN OMELETTE WITH SPAGHETTI, ZUCCHINI AND BACON

意大利煎蛋饼

意大利面 100 克	100g spaghetti
鸡蛋 6 个	6 eggs
奶油 50 克	50g cream
马苏里拉水牛奶酪 100 克，切块	100g buffalo mozzarella, cubed
橄榄油 40 克	40g olive oil
培根 60 克，切碎	60g pancetta, cubed
西葫芦 200 克，切片	200g zucchini, sliced
盐适量	Salt, to taste
黑胡椒适量	Black pepper, to taste

制　　作　将意大利面煮至有嚼劲的程度，捞出后放在盘中，加入适量橄榄油拌匀直至冷却，这样它们就不会粘在一起。

烤箱预热至 200℃。把鸡蛋打在一个足够大的碗里，加入盐、黑胡椒、马苏里拉水牛奶酪块和奶油。

小火加热平底锅，倒入橄榄油，油热之后加入培根碎煸炒，直到变脆。

之后加入煮好的意大利面和西葫芦片，翻炒 1 分钟，再把鸡蛋混合物倒在意大利面上，小火加热 3~4 分钟后，将煎蛋饼翻面，转大火加热 1 分钟，最后放入预热好的烤箱内烤 5 分钟。

把煎蛋饼放入盘子里，静置几分钟后就可以上桌了。

Prepare the dish

Boil the spaghetti until al dente. Drain and leave to cool on a tray, mixing with a little olive oil so the spaghetti doesn't stick together.

Preheat the oven to 200℃. Crack the eggs into a large bowl and add the salt, black pepper, the buffalo mozzarella, cut into small chunks, and the cream.

Heat the skillet over a gentle heat and add the olive oil. When the skillet is hot, add the pancetta and allow it to crisp.

Now add the cooked spaghetti and the zucchini slices. Let them roast for a minute in the fat, then pour the egg mixture on top of the pasta. Let the frittata cook gently for 3~4 minutes and then flip. Set for 1 minute over a high heat and then finish in a hot oven for another 5 minutes.

Transfer the frittata from the skillet to a plate. Leave it to rest for a few minutes and then serve.

ASPARAGI BIANCHI E CAPESANTE
WHITE ASPARAGUS SOUP WITH
SEARED SCALLOPS

白芦笋汤配香煎北海道扇贝

橄榄油 80 克	80g olive oil
大葱 200 克	200g leeks
牛奶 1 升	1L milk
白芦笋 800 克	800g white asparagus
黄油 80 克	80g butter
北海道扇贝 8 个	8 Japanese scallops
培根 8 片	8 slices of bacon
盐适量	Salt, to taste
白胡椒适量	White pepper, to taste

餐盘来自 LEGLE PORCELAIN（法国丽固）"如意"描金 & 描铂金 Ruyi gold Rim & Platinum Rim 系列。
如意寓意顺心、吉祥，其"方中带圆、圆中有方"的形象，传达着东方美学的和谐气氛。技法纯熟
的工匠以手绘技艺将奢华黄金、铂金绘制到瓷器的边角及握柄处，线条细致匀称，既继承了典雅优
美之古风，又融合了时尚现代之活力。
LEGLE PORCELAIN（法国丽固）1953 年始创于法国利摩日，是法国两大造瓷家族——罗格朗家族和
罗布克家族的结晶。两大家族的历史可追溯到 19 世纪，以其精湛的工艺和鲜艳生动的釉面色彩技艺
著称。

制　作　将白芦笋去皮，切片，白芦笋尖留作装饰用。大葱洗净，只留下葱白，切成细丝。
在一口大锅中，放入橄榄油和大葱丝煸炒，待炒出香味之后，加入适量牛奶。另取锅，用剩余牛奶煮白芦笋，再将牛奶和白芦笋全部放入大锅中，文火慢炖直至芦笋变软，用盐和白胡椒调味。
把锅内所有液体倒入食品料理机，搅打 5 分钟后用细筛过滤。将过滤后的液体再重新放回锅中，加入黄油，用文火炖煮。如果汤太过浓稠，可以加一些水。
用培根把扇贝包裹起来，用盐和白胡椒调味。平底锅中倒入少许橄榄油，将培根煎至焦脆，扇贝温热。千万不要煎制过度。
在同一口锅中，放入白芦笋尖快速炒一下。
在四个盘子中分别倒入汤，每个盘中放两个培根扇贝卷，用白芦笋尖装饰。

Prepare
the dish　　Peel and chop the white asparagus, reserving the tips. Wash the leeks and julienne the white part. Sweat the leeks in a large pot with some olive oil for a few minutes until fragrant. Then add some milk. Boil the asparagus with the remaining milk in another pot, then put them all into the large pot with the leeks. Cook slowly until the as paragus are soft and season with salt and white pepper.
Transfer the soup to a blender and blend for about 5 minutes. Strain the soup through a sieve. Put the soup back over a low heat and add the butter. If the soup is too thick, add some water. Wrap the scallops in the bacon, season with salt and white pepper. Heat some olive oil in a cast iron pan and cook the scallops until the outside is crispy and the inside is just warm. Be sure not to overcook.
In the same pan, quickly roast the asparagus tips.
Pour the soup onto four plates and garnish with two scallops per plate and the asparagus tips.

LINGUINE CON GRANCHIO E ZUCCHININE

LINGUINE WITH KING CRAB AND

ZUCCHINI

意式扁面配帝王蟹和节瓜

柠檬 1 个	1 lemon
芹菜 100 克	100g celery
欧芹叶 20 克	20g flat leaf parsley
活帝王蟹 1 只	1 live king crab
大蒜 2 瓣	2 garlic cloves
辣椒 1 个	1 chilli
节瓜 200 克，切丝	200g zucchini, julienned
白兰地 100 毫升	100ml brandy
橙皮 5 克	5g orange zest
橄榄油 100 克	100g olive oil
意式扁面 320 克	320g linguine verrigni
乌鱼子① 10 克	10g bottarga
盐适量	Salt, to taste
白胡椒适量	White pepper, to taste
海胆，根据个人口味准备	Sea urchins, to serve

注①：腌渍晒干后的乌鱼子，意大利传统食材。

制　　作	在一口大锅中倒入水，加入对半切开的柠檬，放入芹菜和欧芹叶，煮沸后加入帝王蟹煮 6 分钟。随后将帝王蟹捞出，放入冰水中。 剥出蟹肉，摘下蟹腿，小心去除所有的蟹壳。 在平底锅内放入大蒜、辣椒、节瓜丝，翻炒后加入蟹肉，并淋上少许白兰地，继续翻炒几分钟，让酒液完全挥发。加入橙皮，用盐和白胡椒调味。 将意式扁面在沸水中煮 7 分钟后捞出，与蟹肉、少许煮面水、乌鱼子和橄榄油混合均匀。装盘时可以用蟹腿肉和海胆来点缀。
Prepare the dish	Fill a very large pot with cold water and add the lemon, cut in half, the celery, and the parsley, and bring to the boil. When the water is boiling, add the crab and cook for about 6 minutes. Remove the crab and place in a bath of ice water. Remove the crab meat from the body and the legs, and then carefully remove all the filament and shell. In a sauté pan, heat a little olive oil and add the garlic, the chilli, and the zucchini. Add the crab meat and sprinkle with brandy, then cook for a few minutes to let the alcohol evaporate. Add the orange zest and season well with salt and white pepper. Cook the linguine for about 7 minutes in a large pot of boiling water. Strain the pasta and mix it with the crab, a touch of the cooking water, the bottarga, and the olive oil. Garnish with some pieces of crab leg meat and sea urchins.

多宝鱼柳配茴香及西西里风味茄子酱

意式茄子酱	For the caponata
茄子 500 克	500g eggplants
芹菜 100 克	100g celery
花生 50 克	50g peanuts
白葡萄酒醋 80 克	80g white wine vinegar
白砂糖 40 克	40g sugar
切里尼奥拉橄榄 50 克	50g Cerignola olives
黑胡椒适量	Black pepper, to taste
盐适量	Salt, to taste
橄榄油，视需求定	Olive oil, as necessary

茴香高汤	For the fennel broth
茴香头 1 公斤	1kg fennel bulbs
大葱 100 克	100g leeks
柠檬 1 个	1 lemon
白葡萄酒 50 毫升	50ml white wine
茴香籽 20 克	20g fennel seeds
大蒜适量	Garlic, to taste
多宝鱼柳 800 克	800g turbot fillets
百里香适量	Thyme, to taste
盐适量	Salt, to taste
黑胡椒适量	Black pepper, to taste
橄榄油，视需求定	Olive oil, as necessary

茄 子 酱　　　将茄子和芹菜（梗）洗净，分别切成 1 厘米厚的丁。锅中倒入适量橄榄油，翻炒茄丁直至变软，随后把茄丁放入大碗内，备用。用余下的橄榄油炒芹菜丁，注意时间不要过长。将炒好的茄子丁与芹菜丁混合后备用。

把花生放入烤箱，180℃烤 5 分钟。

把白砂糖和白葡萄酒醋混合后煮沸，直至质地变得浓稠。

橄榄去核、切丁。把橄榄、花生、糖醋汁与茄丁、芹菜丁混合，用木勺搅拌，用盐和黑胡椒调味后，静置备用。

Prepare the caponata

Cut the eggplant into 1cm dice. Clean and cut the celery heart (the tender, white part) into dice the same size as the eggplant. In a large pot, roast the eggplant with some olive oil until soft then transfer to a mixing bowl. Cook the celery in the remaining olive oil. Be sure not to overcook. Add the celery to the eggplant.

Toast the peanuts in the oven for 5 minutes at 180℃.

Bring the white wine vinegar and the sugar to the boil and cook until you get a thick syrup.

Remove the stones from the olives and dice the flesh. Add the olives, the peanuts, and the vinegar syrup to the vegetables and mix them with a wooden spoon. Season with salt and black pepper and keep at room temperature.

茴香高汤　　　茴香头一切为四。柠檬对半切开，大葱切段。

锅中加入适量水，放入茴香头、柠檬、大葱段、白葡萄酒和茴香籽，文火煮制，直至汤汁浓缩至一半。

将汤汁过筛滤掉残渣，用盐和黑胡椒调味。

Prepare the fennel broth

Cut the fennel bulbs into quarters. Cut the lemon in half, and cut the leeks into chunks.

Place the fennel bulbs, the lemon, the leeks, the white wine, and the fennel seeds in a pot with some cold water and bring to a simmer over a low heat. Reduce the liquid by half.

Strain through a sieve and season with salt and black pepper.

多 宝 鱼　　　锅内倒入适量橄榄油，大火将大蒜和百里香翻炒出香味，放入多宝鱼柳，煎至外焦里嫩。

在盘中放上一勺茄子酱，再放上煎好的多宝鱼柳，加入少许茴香高汤，可以用一些应季蔬菜点缀。

Prepare the turbot

Heat the garlic and thyme with some olive oil over a high flame. Add the turbot fillet and roast until well browned on the outside and just cooked on the inside.

Place one spoon of caponata on a plate and top with the roasted turbot fillet. Add some fennel broth, and garnish with seasonal vegetables, if you like.

巧 克 力 复 活 节 彩 蛋

制　　作　　选择一个你喜欢的巧克力蛋模具，打磨光滑后放在一边备用。

选择你最喜欢的巧克力，黑巧克力、牛奶巧克力或白巧克力均可，将其隔水加热至 45℃ ~50℃。不同种类的巧克力需要合适的温度进行融化：黑巧克力为 30℃ ~31℃，牛奶巧克力为 28℃ ~29℃，白巧克力为 26℃ ~28℃。

待巧克力融化后，将巧克力倒进模具中，轻轻晃动它，使巧克力均匀地分布在模具中，然后翻转它，让多余的巧克力滴下来。

确保模具的边缘清洁干净，然后将模具放入冰箱冷藏至少 12 小时。

准备合并巧克力蛋时，加热平盘或平底锅的底部，取出两半巧克力蛋，在锅的表面摩擦几秒钟，然后快速将两半巧克力蛋连接在一起，并保持静止，以便它们连接在一起。

你可以在巧克力蛋中放入任何惊喜，如坚果、糖果、礼物等，但要确保在此过程中始终戴着手套，以免在巧克力外壳上留下指纹。

Prepare
the dish

Choose a chocolate egg mould (any size will work). Polish it well and set it aside.

Choose your favourite type of chocolate and gently melt it to 45℃ ~50℃ in a plastic bag in a water bath. Different types of chocolate will need to be tempered to different final temperatures: 30℃ ~31℃ for dark, 28℃ ~29℃ for milk, and 26℃ ~28℃ for white.

Once the chocolate is ready, grab your mould and pour in the chocolate, moving and shaking the mould to spread the chocolate evenly. Flip the mould upside down and let the excess chocolate drip out.

Make sure to clean up the edges of the mould, and then let the chocolate set for at least 12 hours in a cold room.

When you are ready to form your finished egg, gently warm the bottom of a flat tray or pan, demould the two egg halves and rub the edges on the warm surface for a couple of seconds, then quickly join the two halves together and hold in place for a couple of seconds.

You can place anything you like inside the egg – nuts, candy, gifts and so on – just make sure to always work with gloves so as not to leave fingerprints on the chocolate.

五月

Maggio
May

我特别喜欢五月，此时，万物生长正处于巅峰期，新鲜香草都已经上市了，天气温暖宜人。

I really like May because the weather is perfectly nice and warm and everything is thriving. There are also plenty of fresh herbs available.

在我父亲的家乡阿布鲁佐大区的 Teramo 镇，有一款传统汤肴，是用很多谷物和火腿一起熬制而成的，叫做 "Virtù di Maggio"，它的字面意思就是 "五月的美德"。相传，这款汤是用来进奉给 Maja 女神以祈求丰收的。传说中这款汤的食谱中需要 7 种豆、7 种香草、7 种肉、7 种意面、7 种时令蔬菜，还需要烹饪 7 小时，因为 "7" 代表了人类的美德。如果你在五月来到 Teramo 旅行，你便可以品尝到这道时刻提醒着我要记住意大利祖辈们为了信仰虔诚准备、精心烹饪的著名汤品。

In my father's hometown, Teramo in Abruzzo, there is a traditional soup made with grains and ham trimmings. It is called "Virtù di Maggio", literally meaning "the virtues of May". The soup honours the goddess Maja, who makes the harvest more fertile. The legendary recipe calls for seven legumes, seven herbs, seven meats, seven pasta shapes, and seven seasonal vegetables, and is cooked for seven hours because there are seven human virtues. If you travel to Teramo during this time of year, you should definitely try this well-known dish which my Italian ancesters prepared to show their faith.

五月还有一个重要的日子，5 月 13 日是我的生日。虽然已经很多年没有跟意大利的家人们一起过生日了，但小时候在家过生日的情景依然历历在目。我是家中最小的孩子，我有两个姐姐，她们都特别疼爱我。我的大姐 Paola 做得一手好菜，她最拿手的就是烤蛋糕，直到现在，我跟我太太都认为大姐烤的蛋糕是全世界最好吃的。每年我生日的那天，一家人都要聚在一起，大姐会亲自为我烤一个生日蛋糕，那样的日子真是令人怀念。

May 13th is also an important day because it's my birthday! Although I haven't been able to spend my birthday with my family for a long time, I can still remember how I spent my birthday with them when I was a child. I am the youngest child in my family, and I have two elder sisters who love me very much. My eldest sister Paola is good at cooking, especially baking cakes. Even to this day, my wife and I regard her cakes as the best. Every year, my family would gather together on my birthday, and Paola would make me a birthday cake. I really miss those days.

Menu

金枪鱼、鹌鹑蛋及豌豆沙拉
Insalata Nizzarda
Tuna, Quail Egg and Green Bean Salad

什锦豆意大利火腿汤
Virtú di Maggio
Bean and Ham Soup

意式饺子配菠菜及帕玛森芝士
Ravioli Burrata, spinaci e Parmigiano Reggiano
Burrata Ravioli with Spinach and Parmigiano Reggiano

慢炖鸡肉及混合菌菇
Pollo alla Cacciatora
Braised Chicken with Mixed Mushrooms

那不勒斯 Ricotta 芝士蛋糕
Pastiera Napoletana
Neapolitan Ricotta Cake

金枪鱼、鹌鹑蛋及豌豆沙拉

酱汁
红酒醋 80 克
小葱 50 克
白砂糖 20 克
橄榄油 80 克
大蒜 1 瓣
盐适量
白胡椒适量

沙拉
金枪鱼 400 克
豌豆 500 克
玉兰菜 100 克，切丝
樱桃番茄 100 克，各切四瓣
水煮鹌鹑蛋 4 个，各对半切
鱼子酱 40 克
盐适量
黑胡椒适量

For the dressing
80g red wine vinegar
50g shallots
20g sugar
80g olive oil
1 garlic clove
Salt, to taste
White pepper, to taste

For the salad
400g maguro tuna
500g green beans
100g endives, julienned
100g cherry tomatoes, quartered
4 boiled quail eggs, halved
40g caviar
Salt, to taste
Black pepper, to taste

酱　汁	把所有食材放入搅拌机中搅拌数分钟，然后倒入碗中，放入冰箱冷藏。

Prepare the dressing	Put all the ingredients in a blender and process for a few minutes. Put in a bowl and keep in the fridge.

沙　拉	将金枪鱼切成 2 厘米厚的方块，用盐和黑胡椒腌制后，快速地将每一面煎至变色。 将豌豆放入沸水里煮 1 分钟，然后放入冰水中冷却。 把所有蔬菜放在一个碗里，倒入酱汁均匀搅拌。 将蔬菜摆盘，再放上金枪鱼、鱼子酱和鹌鹑蛋。

Prepare the salad	Cut the tuna into 2cm cubes and then season with salt and black pepper, and then quickly sear each side. Boil the green beans for 1 minute and then plunge into ice water. Place all the vegetables in a large bowl and toss with the dressing. Arrange the vegetables neatly on a plate and top with the tuna cubes, the caviar, and the quail eggs.

餐盘来自 LEGLE PORCELAIN （法国丽固） 戴安娜 Diana 系列。白色代表纯洁，而铂金意味着奢华，图案中不规则的线条好似戴安娜王妃的活泼性格。特别设计此图案来展现英式优雅，以此纪念戴安娜王妃。

什 锦 豆 意 大 利 火 腿 汤

扁豆 50 克	芹菜 50 克，切丁	50g lentils	50g celery, chopped
红腰豆 150 克	胡萝卜 50 克，切丁	150g red beans	50g carrots, chopped
白豆 150 克	帕尔玛火腿 50 克 (片状)	150g white beans	50g Parma ham, in one piece
鹰嘴豆 50 克	佩科里诺干酪 60 克	50g chickpeas	60g Pecorino cheese
托斯卡纳豆 150 克	混合形状意面 230 克	150g Tuscan beans	230g mixed pasta
鼠尾草 10 克	橄榄油 40 克	10g sage	40g olive oil
辣椒 1 个	熟透的番茄 2 个	1 chilli	2 ripe tomatoes
大蒜 2 瓣	盐适量	2 garlic cloves	Salt, to taste
迷迭香 10 克	黑胡椒适量	10g rosemary	Black pepper, to taste

制　　作　　将所有豆子放在水里浸泡 12 小时。

锅中倒入橄榄油，将切好的蔬菜丁、大蒜、辣椒和所有香草放入锅中，中火炒至大蒜呈金黄色后把大蒜和香草挑出。

将泡好的豆子和帕尔玛火腿放入一个玻璃碗，用凉水没过。之后倒入锅中，文火煮至豆子变软，扁豆尤其软烂。将混合意面加入锅中，煮至弹牙。用盐和黑胡椒调味。

将面和汤盛入碗中，撒上佩科里诺干酪，淋上少许橄榄油即可。

Prepare
the dish

Soak all of the dry beans in fresh water for 12 hours.

Heat some of the oil in a large pot over a medium heat and add the chopped vegetables, the garlic, the chilli, and the herbs. Roast until the garlic is golden then remove the garlic and herbs.

Add the beans and the Parma ham in one chunk and cover with cold water. Cook slowly until the beans are soft and the lentils fall apart. Add the mixed pasta and cook until al dente. Season with salt and black pepper.

Finish the soup with Pecorino and some olive oil.

意式饺子配菠菜及帕玛森芝士

新鲜意面面团 400 克	400g fresh pasta dough
菠菜 200 克	200g spinach
意大利乳清奶酪 200 克	200g ricotta
鸡蛋 1 个	1 egg
帕玛森芝士 150 克	150g Parmigiano Reggiano
鸡汤 100 毫升	100ml chicken stock
黄油 100 克	100g butter
盐适量	Salt, to taste
白胡椒适量	White pepper, to taste

制　　作　用压面机将新鲜意面面团压制成薄面片，然后将面片切成数个边长为 4 厘米的正方形面皮。

将菠菜洗净，放入沸水中焯一下，用手将多余的水分挤出，随后切碎。

搅拌机中加入菠菜、意大利乳清奶酪、蛋黄，再加入帕玛森芝士和盐，混合搅拌均匀。将混合后的馅料放入一个裱花袋中。

把馅料挤在面皮中间，用水把面皮的四周浸湿，将面皮折成三角形，再用叉子把面皮的边缘封上。

制作酱汁。在平底锅中加入鸡汤和黄油，文火加热至汤汁浓稠，用白胡椒调味。

另取锅，将意大利饺子在沸水中煮 2 分钟，随后，将饺子捞入酱汁锅中，加少许盐调味。放入帕玛森芝士，再加热 1 分钟即可。

Prepare
the dish

Using a pasta machine, roll out the pasta dough very thinly. Cut the pasta dough into 4cm squares.

Wash and boil the spinach then press with your hands to remove any excess moisture. Chop into small pieces.

In a large mixing bowl, mix the spinach, ricotta, and the egg yolk, add the Parmigiano Reggiano and the salt and mix very well. Put the mixture in a piping bag.

Now, place some of the filling in the centre of each square of pasta. Spray some water on the edges of the pasta squares and fold them in half to get a triangle. Use a fork to seal the edges.

Now prepare the sauce. In a sauté pan, heat the chicken stock and the butter over a gentle heat and reduce gradually. Season with white pepper.

In another pot, boil the ravioli for about 2 minutes and then add them to the sauce, adding salt if needed. Add some more Parmigiano Reggiano and cook for another minute, then plate up.

POLLO ALLA CACCIATORA
BRAISED CHICKEN WITH MIXED
MUSHROOMS

慢炖鸡肉及混合菌菇

干牛肝菌 50 克	50g dried porcini
去骨鸡腿肉 6 个	6 deboned chicken legs
面粉，视需求定	Flour, as necessary
黄油 100 克	100g butter
橄榄油 40 克	40g olive oil
大蒜 2 瓣	2 garlic cloves
迷迭香 2 枝	2 rosemary stalks
白葡萄酒 200 毫升	200ml white wine
芹菜 100 克，切 0.5 厘米小丁	100g celery, cut into 1/2 cm dice
胡萝卜 100 克，切 0.5 厘米小丁	100g carrots, cut into 1/2 cm dice
洋葱 100 克，切 0.5 厘米小丁	100g onions, cut into 1/2 cm dice
蘑菇 800 克	800g mushrooms
草菇 200 克	200g straw mushrooms
鸡汤 500 毫升	500ml chicken stock
盐适量	Salt, to taste
黑胡椒适量	Black pepper, to taste

制　　作　　将干牛肝菌提前放入清水中浸泡 2 小时。

在鸡腿肉上撒一些面粉，用盐和黑胡椒调味。

在铸铁锅中加入一半黄油和一半橄榄油，再加入 1 瓣大蒜和 1 枝迷迭香，中火炒出香味。将鸡腿肉放入锅中，煎至表皮酥脆，再倒入白葡萄酒，直至酒液蒸发。

另起锅放入余下的橄榄油、大蒜和迷迭香，再加入洋葱丁、芹菜丁和胡萝卜丁，略煸炒。

将牛肝菌、蘑菇和草菇洗净切成片，放入蔬菜中，文火略炒，随后加入鸡腿肉。最后倒入鸡汤，煮至软嫩。

Prepare
the dish

2 hours before you start cooking, put the dry porcini in a bowl with fresh water.

Sprinkle the chicken with some flour and season with salt and black pepper.

In a cast iron pan, heat half of the butter and half of the olive oil over a medium flame with 1 garlic clove and 1 rosemary stalk until fragrant. Roast the chicken legs until the skin is crispy, then sprinkle with the white wine and let it evaporate.

Heat the remaining olive oil, garlic, and rosemary in a large pot and add the onion, celery, and carrot.

Clean and slice the porcini, mushrooms and straw mushrooms then add them to the vegetables. Cook gently for a few minutes, and then add the chicken legs. Cover with chicken stock and simmer until tender.

那不勒斯 Ricotta 芝士蛋糕

甜面团
黄油 150 克
冰糖 100 克
盐 2 克
香草荚 1/8 根
柠檬皮（1 个）
蛋黄 40 克
蛋糕粉 250 克

馅料
牛奶 300 克
熟荞麦 450 克
香草荚 1.5 根
柠檬皮（2 个）
黄油 45 克
鸡蛋 180 克
乳清干酪 525 克
白砂糖 390 克
蜜饯橙 150 克

蜂蜜冰激凌
葡萄糖粉 60 克
稳定剂 10 克
白砂糖 155 克
牛奶 1036 克
奶油 370 克
蜂蜜 100 克
香草荚 2 根
蛋黄 70 克
橙花水适量

For the sweet dough
150g butter
100g icing sugar
2g salt
1/8 vanilla pod
Zest of 1 lemon
40g egg yolks
250g cake flour

For the pastiera filling
300g milk
450g cooked buckwheat
1.5 vanilla pods
Zest of 2 lemons
45g butter
180g eggs
525g ricotta
390g sugar
150g candied oranges

For the honey gelato
60g dextrose powder
10g stabiliser
155g sugar
1036g milk
370g cream
100g honey
2 vanilla pods
70g egg yolks
Orange blossom water, to taste

| 甜面团 | 把黄油、冰糖、盐、香草荚、柠檬皮放入碗里搅拌均匀，之后加入蛋黄继续搅拌，直到全部融合在一起，没有黄油块。最后加入蛋糕粉，搅拌成混合面团。将面团用保鲜膜包起来，在冰箱里冷藏几个小时，使用前再取出。 |

| Prepare the sweet dough | Put butter, icing sugar, salt, vanilla, and lemon zest in the bowl of a stand mixer and mix with the paddle attachment. When the batter is homogeneous, add the egg yolks and mix thoroughly until it has been incorporated and there are no lumps of butter. Lastly, add the cake flour and mix just until the dough comes together. Wrap the dough in plastic film and chill it in the fridge for a few hours before using. |

| 馅料 | 将牛奶、熟荞麦、柠檬皮和香草荚一起放入锅中，慢慢煨煮，直到荞麦将牛奶吸收，大约需要20分钟。期间需要一直搅拌，否则混合物会粘在锅底。
加入黄油，再煮一会儿，离火，冷却。
在碗里放入鸡蛋、乳清干酪和白砂糖，使用手动搅拌器混合均匀。
把煮好的荞麦和乳清干酪糊混合在一起，加入蜜饯橙。一直放置到组装时。 |

| Prepare the pastiera filling | Combine the milk with the cooked buckwheat, lemon zest, and vanilla pod in a pot, and slowly simmer them for around 20 minutes until the wheat has absorbed some of the milk. Stir constantly so the mixture doesn't stick to the bottom of the pan.
Add the butter and let the mixture simmer for a little longer. Take off the heat and leave to cool.
In a bowl, whisk the eggs with the ricotta and the sugar, blend with a hand blender.
Combine the cooked buckwheat with the ricotta mixture and add the candied orange. Reserve until ready to assemble the pie. |

| 蜂蜜冰激凌 | 在碗里把干性材料混合在一起。取一口中等大小的平底锅，放入牛奶、奶油、蜂蜜、香草荚和蛋黄，加热，再把干性混合物倒入锅中，煮至约82℃后，将混合物放入罐子里，放入冰箱冷藏至少12小时。第二天取出，搅拌后再放入冰箱里冷冻。 |

| Prepare the honey gelato | Mix the dry ingredients together in a bowl. In a medium saucepan, heat the milk, cream, honey, and vanilla pod and add the egg yolks. Sprinkle the dry ingredients over and cook the base to 82°C. Transfer the mixture to a jug and refrigerate it for at least 12 hours. The next day, churn the gelato and keep it in the freezer. |

组　装　从冰箱中取出甜面团，擀至约 4 毫米薄。取一个直径约 30~35 厘米的圆形蛋糕模，将面团铺在里面。用叉子在面团上戳几个洞，倒入馅料，将多余的面团切下，以便面团与馅料表面保持平齐。在面团表面装饰一些条状的面团（像传统的馅饼），然后将其放入 160℃的烤箱中烤至少 45 分钟。一旦面团变成深金色，馅饼就做好了，馅饼的中间部分虽然弯曲但是会很结实。等到面团在烤箱中完全冷却后再取出。

用一个喷雾瓶，在馅饼上面喷上少许橙花水。

Assembly　Take the dough out of the fridge and roll it to 4mm thickness. Take a round cake tin (30~35cm diameter) and line it with the dough. Use a fork to prick some holes in the dough in the bottom of the tin. Pour in the filling and trim the excess dough on the sides of the tin to almost level with the filling.

Decorate the surface with strips of dough (like a traditional tart) and place it in the oven at 160℃ for at least 45 minutes. The pie is ready once the dough is dark golden and the middle is wobbly but firm. Allow the pastiera to cool completely in the oven before taking it out.

Using a spray bottle, mist some orange blossom water over the surface of the pie.

Summer
夏

六月

六月对厨师来说真的是太美妙了！这个时节，有最好的海鲜：牡蛎、扇贝、白鱼等，所以 Ligurian 海鲜汤是这个季节我最钟爱的美食之一。由于日照越来越充足，罗勒还有其他香草都在疯狂地生长，番茄的味道尤为浓郁，再配上清香的水牛乳酪，在夏天的意大利当然要吃跟意大利国旗相同颜色的绿白红三色 Caprese 沙拉啦。意大利的六月是完美的，你可以尽情地享受海边的微风和日光浴。

June is a great month for chefs! You can enjoy the best seafood such as oysters, scallops, and white fish, in June. Ligurian seafood soup is one of my favourites in this season. Thanks to the abundant sunshine, basil is at its best and tomatoes are ripe as well. If you combine them with fresh buffalo mozzarella, you can enjoy a Caprese salad which has the same colour as the Italian flag. June is also the perfect month to sunbathe and enjoy the sea breeze on the coast.

六月对意大利的孩子们来说意味着学期结束，假期开始啦！当我还是孩子的时候，每年六月，我们举家会从贝加莫搬到阿布鲁索海边的度假屋去度过夏天。记得我父亲的家乡靠近山区，即使不靠近海边，每天清晨我们都会乘短途巴士前往亚得里亚海名为 Roseto Degli Abruzzi 的美丽的白色沙滩游玩。在海边小城的日子非常悠闲，过不了多久人们就要搞出什么意面节、海鲜汤节之类的节日，总之一切的庆祝都跟食物有关。那时候，我特别喜欢去爷爷奶奶家，每周奶奶都会给我们做一次好吃的"妈妈面"，鲜美的番茄浓酱配上手工自制的小肉丸，这道菜便是我从事厨师这个职业的初心所在。

June is a great month for kids in Italy — school is finished and it is holiday time! When I was younger, my family would drive from Bergamo to Abruzzo to spend the summer there. Every morning we would take a short bus drive to Roseto Degli Abruzzi, a beautiful white sandy beach on the Adriatic Sea to lead a leisurely life by the sea. We would go to pasta festivals and seafood festivals from time to time - all the celebrations were always about food! We would visit my grandparents and enjoy some great food. Every week, my grandma would make her special meatball pasta with tomato sauce, the dish that inspired me to become a chef.

另外，6 月 2 日是意大利国庆节，在罗马每年都会举办盛大的游行，小时候我们全家都会聚在一起看国庆节的游行直播，那情形就像中国人过年回家看"春晚"一样。

June 2nd is Republic Day in Italy and our whole family would get together when I was a kid to watch the live broadcast of the parades happening in Rome every year. It's just like when Chinese people go home for the Spring Festival Gala.

Menu

意式经典沙拉配新鲜番茄及水牛奶酪
Insalata Caprese
Tomato and Buffalo Mozzarella Salad

利古里亚风味海鲜汤
Caciucco alla Livornese
Ligurian-style Seafood Soup

手工意式细面配小肉丸及番茄汁
Spaghetti alla Teramana
Homemade Spaghetti, Veal and Pork Meatballs

低温慢炖小牛胸及时令蔬菜
Cima alla Genovese
Simmered Veal Brisket Stuffed with Seasonal Vegetables

椰子
Cocco
Coconut

INSALATA CAPRESE
TOMATO AND BUFFALO MOZZARELLA
SALAD

意式经典沙拉配新鲜番茄及水牛奶酪

罗马番茄 4 个	4 roma tomatoes
水牛奶酪 400 克	400g buffalo mozzarella
橄榄油 80 克	80g olive oil
新鲜罗勒叶 20 克	20g fresh basil leaves
樱桃番茄 100 克	100g cherry tomatoes
罗勒汁 100 克	100g basil dressing
盐适量	Salt, to taste
黑胡椒适量	Black pepper, to taste

制　作
将罗马番茄放在沸水里烫 20 秒后去皮。
将罗马番茄和水牛奶酪切成 0.5 厘米厚的片，用盐和黑胡椒调味，并刷上橄榄油。
盘中放入番茄片和水牛奶酪片。用罗勒叶和樱桃番茄点缀，再淋上少许罗勒汁，最后撒少许盐即可。

Prepare the dish
Blanch the roma tomatoes for 20 seconds and remove the skin.
Cut the tomatoes and the buffalo mozzarella into 0.5 cm-thick slices. Season well with salt and black pepper, and brush with olive oil.
On a serving plate, alternate the tomatoes and buffalo mozzarella slices. Arrange some basil leaves, cherry tomatoes, and basil dressing on the plate, and finish with salt.

LEGLE
PORCELAIN

餐盘来自 LEGLE PORCELAIN （法国丽固）地中海 Mediterranean 系列。文艺复兴前的西欧，家居艺术经过浩劫与长时期的萧条后，在 9 至 11 世纪又重新兴起，并形成独特的风格——地中海式风格。地中海风格是最富有人文精和艺术气质的风格之一。

CACIUCCO ALLA LIVORNESE
LIGURIA-STYLE SEAFOOD SOUP

利古里亚风味海鲜汤

橄榄油适量	Olive oil, to taste
辣椒 1 个	1 chilli
百里香适量	Thyme, to taste
鼠尾草 30 克	30g sage
大蒜 2 瓣	2 garlic cloves
螃蟹 500 克	500g crab
墨鱼仔 100 克	100g baby calamari
马头鲷 200 克	200g john dory
海虹 300 克	300g mussels
番茄 400 克	400g tomatoes
洋葱 100 克，切碎	100g onions, chopped
鳕鱼 200 克	200g cod
鮟鱇鱼 200 克	200g monkfish
红葡萄酒 300 毫升	300ml red wine
鳌虾 6 个	6 scampi
托斯卡纳蒜香面包 1 片	Tuscan bread, one piece

制　　作　　在铸铁锅中加入橄榄油、辣椒、百里香、鼠尾草和大蒜，中火煸炒，再放入墨鱼仔和螃蟹，使其入味。

把马头鲷切块，加入锅中，翻炒几分钟后加入红葡萄酒，直至酒液蒸发。在锅中加入番茄，熬煮20 分钟。

另取一口小锅，加入橄榄油，放入洋葱碎和 1 瓣大蒜煸炒，再加入鳕鱼和鮟鱇鱼，加水没过鱼，熬煮 20 分钟。

将小锅内的食材全部倒进铸铁锅中，加入海虹，直至煮沸。出锅前 10 分钟加入鳌虾。

将汤盛出，配上一片托斯卡纳蒜香面包，淋上少许橄榄油即可。

Prepare the dish

Heat the olive oil, chilli, thyme, sage, and 1 garlic clove in a cast iron pot over a medium heat. Add the baby calamari and the crab and let the flavours infuse.

Cut the john dory into pieces and add to the pot, then after a few minutes add the red wine and let it evaporate. Add the tomatoes and cook the mixture for about 20 minutes.

In a smaller pot, heat the olive oil with the remaining garlic clove and chopped onions, then add the cod and the monkfish, cover with water, and cook for 20 minutes.

Now, add the mussels and transfer everything to a large pot and bring to the boil. 10 minutes before you want to serve, add the scampi.

Divide the soup between bowls and serve with a slice of toasted Tuscan bread, drizzle with a little olive oil.

SPAGHETTI ALLA TERAMANA
HOMEMADE SPAGHETTI, VEAL AND PORK
MEATBALLS

手工意式细面配小肉丸及番茄汁

小牛肩颈肉 100 克	100g veal shoulder
猪颈肉 200 克	200g pork neck
帕玛森芝士 100 克	100g Parmigiano Reggiano
鸡蛋 2 个	2 eggs
特级初榨橄榄油 100 克	100g extra virgin olive oil
白洋葱 1 个，切丝	1 white onion, diced
去皮番茄 11 个	11 peeled tomatoes
手工意式细面 400 克	400g Spaghetti
陈年佩科里诺干酪 40 克	40g aged Pecorino
盐适量	Salt, to taste
黑胡椒适量	Black pepper, to taste

肉　丸

将小牛肩颈肉和猪颈肉分别剁成肉馅（手动或用绞馅机），混合后放在大碗里。加入帕玛森芝士、鸡蛋，用盐和黑胡椒调味。混合均匀后，放入冰箱冷藏 2 个小时。随后取出，做成小肉丸，再放回冰箱冷藏。

Prepare the meatballs

Mince the veal shoulder and pork neck (by hand or in a meat grinder) and place the mixture in a large bowl. Add the Parmigiano Reggiano and the eggs, and season well with salt and black pepper. Mix well, then rest in the fridge for two hours. Roll into tiny meatballs and keep refrigerated.

酱　汁

锅中倒入 25 克橄榄油，将洋葱丝煸炒 10 分钟。当洋葱变软后，加入去皮番茄，文火熬煮至少 2 个小时。将熬煮好的酱汁用料理机搅拌均匀。

另取一口平底锅，锅热后倒入剩余 75 克橄榄油，放入肉丸，当肉丸煎至表面焦黄时，倒入酱汁。将意面煮熟，倒上酱汁，再撒上适量陈年佩科里诺干酪即可。

Prepare the sauce

Cook the onion in a large pot with 25g olive oil for at least 10 minutes. When the onion is softened, add the peeled tomatoes and cook slowly for at least two hours. Purée the sauce in a blender. Heat with 75g remaining olive oil in a sauté pan and cook the meatballs. When they are golden brown, remove from the heat and add the sauce.

Cook the spaghetti and when ready, mix with the sauce and finish with a grating of aged Pecorino.

餐盘来自 LEGLE PORCELAIN（法国丽固）地中海系列 Mediterranee
系列。把地中海的阳光带到餐桌上来！白色村庄与沙滩、碧海蓝天连成一片，蔚蓝色的浪漫情怀，谱写出艳阳高照下的纯美自然。

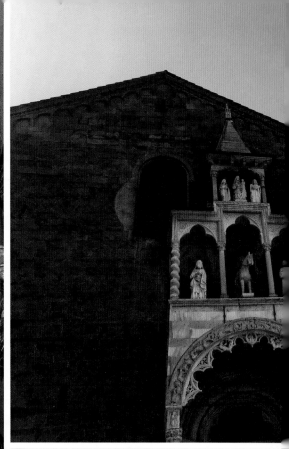

妈妈的故事

MY MOTHER'S STORY

我最爱的大餐是我妈妈做的菜，是她教会我如何去尊重每一种食材，事实上任何一款小叶菜，甚至任何一棵香草，都会对烹饪一道完美菜肴产生至关重要的作用。我妈妈的名字叫 Rosanna（罗莎娜），是美丽的玫瑰的意思。她 1941 年 6 月 22 日出生在意大利贝加莫省，她现在依然住在贝加莫省的 Cisano Bergamasco 市郊。如今，她的记忆力出了些问题，很多事情她都不太记得了，不过她依然很可爱。她特别喜欢煮咖啡，在家的时候，每隔 20 分钟她都会问我姐姐 Paola 和姐夫 Alessandro："亲爱的，要不要来一杯咖啡？"

My favourite chef is my mother and she taught me how to respect every ingredient, like how the smallest sprinkling of herbs contributes to a perfect dish. My mum Rosanna was born in Bergamo on June 22nd, 1941, and lived in Cisano Bergamasco. Today she still lives in the suburb of Cisano Bergamasco in Bergamo Province. Her name means beautiful rose. Although she struggles with her memory nowadays, she is still lovely and likes making coffee. When she is at home with my sister Paola and Alessandro, Paola's husband, she will ask them "Beviamo un bel caffé?" every 20 minutes.

时光追溯到 20 世纪 80 年代，那时我还是个孩子，记得每到星期天，一大家子人都会在家中聚会，我的妈妈和姥姥 Nonna Letizia 早早就开始为家人准备晚餐了。Polenta 是贝加莫的传统美食，有点像中国北方的"棒子面粥"，不同的是，我们在粥里要放很多食材：不同种类的红肉、禽类和蔬菜，非常美味！小时候，每个星期天聚会的主题都是 Polenta，这也很像在 20 世纪 80 年代的中国北方，每个星期天一家人聚在一起包饺子，这些记忆实在是太美好了。

Back in the early 1980s when I was a kid, Sunday was a day of celebration in my home. Back then, my grandmother Nonna Letizia and my mother would prepare dinner for the whole family starting early in the morning. Polenta is a traditional dish that is like a Bergamo version of the "corn porridge" in northern China. However, we tend to put more ingredients in it such as different kinds of red meat, poultry, and vegetables. It's delicious! At that time, we would have Polenta every Sunday. It was just like in northern China in the 1980s, when families would get together to enjoy home-made dumplings. The memory is so sweet!

不过，我的父亲 Peppino 出生成长在意大利南部位于阿布鲁索大区中部的小镇 Basciano，他就吃不惯 Polenta，至少他无法忍受每周日都要吃，所以他一有机会就带着妈妈去南部度假，这样他就不用每周都吃 Polenta 了，还能吃些跟北方不一样的美食。我想，味蕾上的"南北之争"，中国人和意大利人都一样，这就好比中国的南方人无法理解为什么在北方，几乎每个节庆日、每个周末都要吃饺子吧。

My Father Peppino was born in Basciano, a town in the middle of Abruzzo in Southern Italy. He didn't think polenta was wonderful, or at least he couldn't stand having it every Sunday. So, if he got the chance, he would take my mum to the south on vacation in order to escape the weekly Polenta. In this way, he could try different food from the north. I think Chinese people are the same as Italians when it comes to the difference in taste between north and south. People in southern China cannot understand why people in the north like to eat dumplings almost every festival and weekend.

同时，我妈妈为了既不改变星期天吃 Polenta 的传统，又要照顾到爸爸的胃口，她就开始创新一些南北口味结合的新菜肴，她对家人满满的爱化作了她对烹饪的热情。每到不同节庆，比如春天的复活节或冬天的圣诞节，她都会根据时令和节日传统，精心为家人烹饪美食。我最爱的就是她做的意大利手工混合肉丸吉他面，这是我妈妈的独创，后来我把这道菜起名为"妈妈面"。

And that's how the situation changed. Sunday was still polenta day, but in addition, my mum started to prepare some new dishes that blended northern and southern flavours. Her passion for cooking would really come out on festive occasions like Easter and Christmas. My favourite dish she makes is spaghetti alla chitarra with handmade meatballs.

如今，每当我在创意新菜品的时候，我都会想起妈妈为家人烹饪一大桌子饭菜的样子，为爱而烹饪，这个场景一直激励着我，妈妈是我的动力源泉。记得 2010 年，我邀请妈妈来到北京，为我当时工作的餐厅 Sureno 客座烹饪，我们一起在厨房工作了一个星期，和妈妈一起做饭的那段时光太珍贵太开心了！很多朋友都专程来餐厅吃我妈妈亲自烹饪的这道"妈妈面"，大家都赞不绝口。之后，我工作过的每家餐厅，菜单上都会有这道"妈妈面"，这道菜代表了我对妈妈的爱。

These days, when I create new dishes in the kitchen, I sometimes think of her and the way she always cooked with great love and dedication. Her passion continues to inspire me. In 2010, I invited her to Beijing and we spent the whole week cooking together at Sureno. I really enjoyed and cherished the time cooking with my mom. My friends still remember Rosanna and her awesome spaghetti alla chitarra — many people came especially to try her cooking! I have taken that spaghetti to every restaurant I have worked at since, showing my love for my mother.

老马的家人

文 / 张际星

如果你见过了老马的家人，你便知道老马的重情重义源自何处。

2018 年初夏，我一个人在米兰出差，突然收到老马发来的微信，他说米兰离他的家乡贝加莫很近，邀请我周末去他妈妈家做客，他妈妈听说我在米兰，跟他说一定让我来家里吃饭。我有点犹豫，毕竟跟他家人只在北京见过一面，老马和他太太都在北京，我自己跑去会不会给他家人添麻烦呀？然而，老马一再坚持，我不好推辞，便答应了。

从米兰到贝加莫只要坐一个小时的火车，到站时，老马的二姐 Sara 已经在站台等候了。Sara 很漂亮，一头金发，总是笑眯眯的，且英文很好。她先带我参观了贝加莫老城区，老城区建造在山顶，四周都是山，视野非常开阔，山景也很漂亮。因为是周末，老城区很热闹，当地人和外来游客都不少，走在古老的石板路上，有点穿越到中世纪的感觉。老马的妈妈家在城郊，从老城区驱车 40 分钟就到了，她家房子很大，还有一个大院子。我们到时，一大家子人都已经做好午饭等我们了。老马的妈妈一见到我就给了我一个大大的拥抱，还在我脸上狠狠地亲了一下，不停地感谢我在北京关照她的儿子，说着说着眼圈就红了，看得出来妈妈很想念老马。

老马的大姐 Paola 做得一手好菜，一大桌子菜都是她在上午一个人做的：四五款特别新鲜的沙拉和三款

不同味道的意面，还有烤牛排和烤鸡，十几款面包和奶酪拼盘，味道绝对不输任何一家米其林餐厅！

我有些不好意思，感谢大姐忙活了一上午，Paola 说如果我不来，周末全家人也都是要聚在一起吃饭的，她和二姐家都离妈妈家很近。这时，Sara 开玩笑说："家里大姐做饭最好吃，我们家最不会做饭的就是老马，哈哈哈哈。"

吃完饭，妈妈端上了咖啡，还特意让我在咖啡里加了一小杯烈酒，说是可以帮助消化，紧接着妈妈自己就喝掉了一杯加了双倍量烈酒的咖啡，非常豪迈。最重磅的甜品来啦！老马曾经在《来吃意大利菜》一书中提到大姐 Paola 是世界上最会烤蛋糕的人，我一直很好奇 Paola 烤的蛋糕究竟有多好吃。

Paola 特地为我烤了三个蛋糕——香草、开心果和巧克力口味的，果然每一个都好吃到停不下来，我能吃到书中极荐的美味，由衷地满足。这时，妈妈又端着咖啡"套餐"走过来，笑眯眯地问我："宝贝，要不要来一杯咖啡？"这时全家人都笑了，原来妈妈年纪大了，记忆力出了些问题，有时候会重复做一些事情，医生也没有什么特别有效的治疗方案。家里人已经习惯了，觉得只要妈妈身体健康，能吃能喝就不是什么大问题，反而觉得她这样也很可爱。

我被老马一家人的乐观、好客和亲情深深感动着，一个人在欧洲停留多天，没想到在贝加莫竟然感受到了小时候一大家人其乐融融的温暖。

低温慢炖小牛胸及时令蔬菜

小牛胸 400 克	400g veal brisket
牛奶 200 毫升	200ml milk
大蒜 2 瓣，切末	2 garlic cloves, crushed
面包碎 100 克	100g breadcrumbs
芹菜 50 克，切碎	50g celery, finely chopped
胡萝卜 50 克，切碎	50g carrots, finely chopped
鸡胸肉 100 克，切丁	100g chicken breast, diced
豌豆 50 克，切丁	50g green beans, diced
橄榄油适量	Olive oil, to taste
柠檬汁适量	Lemon juice, to taste
蛋黄酱适量	Mayonnaise, to taste
盐适量	Salt, to taste
黑胡椒适量	Black pepper, to taste

| 制　　作 | 用小刀沿着小牛胸脂肪的部位，将牛肉切成 1.5 厘米厚的片。用盐和黑胡椒腌制，放入冰箱冷藏。 |

用小刀沿着小牛胸脂肪的部位，将牛肉切成 1.5 厘米厚的片。用盐和黑胡椒腌制，放入冰箱冷藏。

与此同时准备馅料。把大蒜末和牛奶一起煮沸，离火，加入面包碎，用盐和黑胡椒调味，加入芹菜碎、胡萝卜碎、鸡胸肉丁、豌豆丁，搅拌均匀。将混合馅料放在牛肉片上，将牛肉卷起，用线将两端系紧并密封好。

锅中加入凉水，放入牛肉卷煮一个半小时。随后静置到室温，用重物压在牛肉卷上，放入冰箱冷藏一晚。

将牛肉切成薄片摆放在盘子中，淋上橄榄油、柠檬汁或蛋黄酱。

Prepare the dish

Using a paring knife, open the brisket along the fat part and try to create an even thickness of about 1.5 cm. Season the veal brisket with salt and black pepper and put it in the refrigerator.

In the meantime, prepare the stuffing. Bring the milk and the crushed garlic to the boil, then remove from the heat and add the breadcrumbs. Season with salt and black pepper. Add the chopped celery and carrot, the chicken breast, and the green beans. Mix well and spread the stuffing down the centre of the veal brisket. Roll up the veal brisket and tie the ends tightly with kitchen string. Place in a large pot with cold water and simmer for 1.5 hours, then remove from the heat and cool until you can handle it. Place the veal brisket in the fridge overnight with a weight on top.

Thinly slice the veal brisket and serve with olive oil and lemon juice or mayonnaise.

椰 子

巧克力奶油酥饼
黄油 180 克
白砂糖 120 克
80% 黑巧克力 155 克
水 100 克
蛋糕粉 460 克

For the chocolate sable
180g butter
120g sugar
155g 80% dark chocolate
100g water
460g cake flour

椰子甘纳许
椰子泥 (A) 350 克
葡萄糖浆 60 克
转化糖浆 60 克
白巧克力 580 克
淡奶油 1000 克
椰子泥 (B) 350 克
椰子朗姆酒 60 克

For the coconut whipped ganache
350g coconut puree (A)
60g glucose syrup
60g Trimoline
580g white chocolate
1kg whipping cream
350g coconut puree (B)
60g Malibu

黑巧克力甘纳许
淡奶油 (A) 600 克
香草荚 1 根
葡萄糖浆 70 克
转化糖浆 70 克
70% 黑巧克力 475 克
淡奶油 (B) 1200 克

For the dark chocolate whipped ganache
600g whipping cream (A)
1 vanilla pod
70g glucose syrup
70g Trimoline
475g 70% dark chocolate
1.2kg whipping cream (B)

巧 克 力 奶油酥饼	将黄油、白砂糖和黑巧克力混合融化，然后加入水，混合均匀后倒入搅拌机中搅拌，随后慢慢加入蛋糕粉继续搅拌，令其充分混合。 把面团放在冰箱里冷藏。
Prepare the chocolate sable	Melt the butter with sugar and the dark chocolate, mix them and then add the water. Transfer the mixture to the bowl of a stand mixer. While mixing them with a paddle attachment, add the cake flour slowly and mix just enough to combine. Reserve the sable mixture in the fridge until you are ready to make your tart.
椰 子 甘 纳 许	在平底锅中加热椰子泥 (A)，加入葡萄糖浆和转化糖浆，直到混合物融化。在微波炉中融化白巧克力，然后将上述混合物倒在白巧克力上面并混合成乳液状。 一旦混合均匀，加入淡奶油并使用搅拌机混合，最后加入椰子泥 (B) 和椰子朗姆酒，即为椰子甘纳许。将制作好的椰子甘纳许放在冰箱里冷藏。
Prepare the coconut whipped ganache	In a sauté pan, warm the first portion of coconut puree (A) with the glucose and Trimoline until the mixture melts. In a microwave, melt the white chocolate and then pour the first mixture over it and mix to create an emulsion. Once all the ingredients have been absorbed and the mixture is uniform, add the cold cream while mixing with the help of an immersion blender. Then add the second portion of coconut puree (B) plus the Malibu. Reserve the coconut whipped ganache in the fridge until ready to use it.
黑巧克力 甘 纳 许	在平底锅中加热淡奶油 (A)，加入香草荚、葡萄糖浆和转化糖浆，直到混合物融化。在微波炉中融化黑巧克力，然后将上述混合物倒在黑巧克力上面并混合成乳液状。一旦混合均匀，加入淡奶油 (B) 并使用搅拌机混合，即为黑巧克力甘纳许。将制作好的黑巧克力甘纳许放在冰箱里冷藏。
Prepare the dark chocolate whipped ganache	In a sauté pan, warm the first portion of cream (A) with the vanilla, glucose and Trimoline just enough to melt the sugars inside. In a microwave, melt the dark chocolate and then pour the first mixture over it and mix to create an emulsion. Once all the ingredients have been absorbed and the mixture is uniform, add the second portion of cream (B) while mixing with the help of an immersion blender. Reserve the dark chocolate whipped ganache in the fridge until ready to use it.

烘　焙 及 组 合	在您准备烤酥饼的那天，将酥饼面团擀成 2.5 毫米薄，然后折叠面团外圈。 根据酥饼的大小，放入烤箱 160℃烘烤约 20 分钟，要确保酥饼完全烤熟且酥脆。把酥饼放在架子上冷却。 用搅拌机分别搅拌椰子甘纳许和黑巧克力甘纳许，然后用黑巧克力甘纳许填满酥饼的底部，将表面抹平，然后随意在上面用椰子甘纳许做点缀，并用黑巧克力甘纳许填充空隙 。
Baking and assembly	The day you are ready to bake the tart, roll out the chocolate sable to a thickness of 2.5mm then use to line a tart ring. Bake at 160℃ for around 20 minutes depending on the size of your tart, making sure that the tart is fully baked and crispy. Set the tart shell aside on a rack and let it cool down. With the help of a stand mixer, separately whip the two ganaches and then proceed to fill the bottom of the shell with the dark chocolate whipped ganache. Level out the surface and then pipe some spikes with the coconut whipped ganache and fill the gaps with the dark chocolate whipped ganache.

七月

Luglio
July

天气炎热，夏天到来了，意大利人早已按捺不住纷纷奔向海边避暑啦！七月的海边最美，湛蓝的海水清澈见底，随便往沙滩上一躺，晒热了就躲去海里游个来回，日子慢悠悠懒洋洋的，一天一晃就过去了。意大利人最爱吃海鲜，海边的各家餐厅都会在这个月供应各式海鲜大餐。我最喜欢的就是炸海鲜，无论是鱿鱼、大虾、小银鱼，还是一些应季时蔬，一律炸得香酥脆嫩，配上厨师特调的酱汁，再配上一杯酒，怎么吃都吃不够。

As soon as summer arrives, all Italians love to hit the beach to escape the fierce hot weather. The crystal blue waters of the seaside are most beautiful in July. You can just lie on the beach, or swim if you get hot. The lazy days pass so quickly. As for the food, seafood is prefered at this time. Restaurants along the coast serve all kinds of seafood dishes. I like fried squid, prawns, white bait, and seasonal vegetables. They're all crispy, and I can never get enough of them with a chef-made sauce and a chilled glass of white wine.

意大利最美丽的地方之一就是艾马尔菲海岸（Amalfi Coast），这是一个神奇的地方，站在悬崖的观景台上眺望大海，景色令人叹为观止。从观景台远眺，你还能看见一排排游艇停泊在海边，其中很多都是来艾马尔菲或卡普里岛（Capri）游玩的名人或明星的私家游艇。

One of the most beautiful places in Italy is the Amalfi Coast. It is magical. The view from the top of the cliffs is breathtaking. From the terraced villages stacked high above the sea, you can watch the sail boats and the celebrity yachts approaching Amalfi or Capri.

七月还有一件非常棒的事情——无论是比萨饼还是海鲜，都有一种特别的味道！最惬意的事情就是坐在海边或者任何一间靠海的咖啡馆，点上一份"Baba"（用朗姆酒泡制的甜面包），会让你的心情一整天都愉悦绽放！

Another great thing about July is the food. From pizza to seafood, every dish has its own special flavour. One the most best treats is a rum baba. Whether you have it on the beach or in a nice café, it will definitely make your day brighter.

Menu

法国诺曼底蓝龙虾沙拉
Carpaccio di Astice
Normandy Lobster Salad

香脆什锦炸海鲜
Fritto Misto
Mixed Fried Seafood

意式手工通心粉配鳌虾及鸡心豆
Maccheroncini scampi e ceci
Homemade Maccheroncini with Scampi and Chickpeas

烤石斑鱼配新鲜番茄及橄榄
Cernia all'acqua pazza
Oven-Baked Grouper with Tomato and Olives

朗姆巴巴
Rum Baba

CARPACCIO DI ASTICE
NORMANDY LOBSTER SALAD

法国诺曼底蓝龙虾沙拉

龙虾 1 只	1 lobster
柠檬汁 50 毫升	50ml lemon juice
特级初榨橄榄油 100 克	100g extra virgin olive oil
节瓜 200 克	200g baby zucchini
红酒醋 20 克	20g red wine vinegar
橙子酱汁 100 克 (详见一月食谱)	100g orange dressing (see January recipes)
奥西特拉鱼子酱 50 克	50g Oscietra caviar
海盐适量	Sea salt, to taste
黑胡椒适量	Black pepper, to taste
茴香汁适量	Fennel jelly, to taste

制　　作　　将龙虾放入沸水中煮 3 分钟，捞出后放入冰水中冷却。去除龙虾壳及内脏，将龙虾尾及龙虾钳中的肉取出，切成薄片，用柠檬汁、橄榄油、海盐和黑胡椒调味。

将节瓜切片，用橄榄油、海盐、黑胡椒和红酒醋调味。

在盘中放入调好味的节瓜片，将龙虾片摆放在上面，再淋上适量橙子酱汁，舀一勺鱼子酱放在龙虾上，用茴香汁点缀。

Prepare
the dish

Boil the lobster for 3 minutes and then place it in a bath of iced water. Remove the lobster shell and the intestines. Slice the lobster meat taken from the tails and claws, then season with lemon juice, olive oil, sea salt, and black pepper.

Sliced the baby zucchini and season with olive oil, sea salt, black pepper, and red wine vinegar.

On a serving plate, arrange the zucchini salad and cover it with the lobster slices. Add some orange dressing and a large spoon of caviar on top of the lobster. Garnish with fennel jelly.

香脆什锦炸海鲜

海鲜及酱汁

银鱼 400 克

鱿鱼 200 克

扇贝 400 克

中等大小的虾 300 克

橄榄油 20 克

辣椒 2 个

大蒜 2 瓣

番茄酱 100 克

蛋黄酱 100 克

小葱 1 小段

腌渍水瓜柳 40 克

红椒 50 克

柠檬 2 个（1 个挤汁）

油（油炸用）2 升

面糊

面粉 200 克

盐适量

小苏打适量

蛋白 1 个

橄榄油 20 克

冰气泡水 25 毫升

Seafood and sauces

400g white bait

200g calamari

400g scallops

300g medium-sized shrimps

20g olive oil

2 chillies

2 garlic cloves

100g tomato sauce

100g mayonnaise

1 small shallot

40g capers in vinegar

50g red peppers

2 lemons, 1 juiced

2L oil, for deep frying

For the batter

200g flour

Salt, to taste

Baking soda, to taste

1 egg white

20g olive oil

25ml cold sparkling water

| 海　　鲜 | 将银鱼洗净。将鱿鱼头和身体切开，再切片，放入冰箱冷藏。扇贝切片。虾去壳，如果你喜欢，可保留虾尾。 |

| Prepare the seafood | Clean and prepare the white bait. Slice the calamari, separate the heads from the bodies, and keep in the fridge. Cut the scallops into slices. Remove the shell from the shrimps, keeping the tail on, if you like. |

| 酱　　汁 | 辣味番茄酱：在平底锅中加入橄榄油、辣椒和大蒜，再加入番茄酱，熬煮 30 分钟，静置冷却。
水瓜柳酱：把小葱和红椒切成小丁，加入柠檬汁（1 个）、水瓜柳和蛋黄酱，混合均匀。 |

| Prepare the sauces | For the spicy tomato sauce, heat the olive oil in a pan with the chillies and garlic, add the tomato sauce, and cook for 30 minutes. Serve cold.
For the caper sauce, chop the shallots and the red pepper into fine brunoise, then mix together with the lemon juice, the capers, and the mayonnaise. |

| 海鲜制作 | 制作面糊：把面粉、盐和小苏打放在碗内，再加入蛋白、橄榄油、冰气泡水，用搅拌机搅拌 1 分钟。
炸锅内放油，加热至 180℃ ~210℃。将鱿鱼、扇贝、虾、银鱼裹上面糊后逐一放入锅中炸至金黄。炸好后捞出，放在吸油纸上沥油。
将海鲜放入盘中，用柠檬和两种酱汁点缀。 |

| Prepare the batter and fry the seafood | Place the flour, salt, and baking soda in a large bowl. Add the other ingredients, and then mix well for a minute with a blender.
Heat the oil to 180℃ ~ 210℃. Cook the seafood in the following order, passing them quickly through the batter and then frying until golden brown: calamari, scallops, shrimp, and white bait. Once the seafood is cooked, place it on some kitchen paper to absorb any excess oil.
Plate the seafood and garnish with lemon and the two sauces. |

意大利托斯卡纳多菲酒庄
天地匠心 春秋佳酿

温润气候

多菲酒庄坐落在意大利最负盛名的葡萄酒产区托斯卡纳大区，两片共 10 公顷的葡萄园依比萨丘陵的斜坡而立，既有利于排水，又不易积聚冷空气，光照条件也优于平地，是栽培葡萄的最佳地块。周围环绕着一大片已经设立了几十年的自然保护区，这些茂密的树林又起到了调节微气候的作用。主葡萄园 Poggio di Galleta 在 1869 年已有记载，最早属于中世纪古堡 Lari 小城中的贵族所有。

托斯卡纳大区是典型的地中海气候，夏季在副热带高压控制下，气流下沉，气候炎热干燥少雨，云量稀少，阳光充足，这也是大名鼎鼎的"托斯卡纳艳阳"的来历。冬季受西风带控制，气候温和，最冷月份的气温在 4℃ ~10℃ 之间，降水量丰沛。

土壤条件

Giacomo Tachis 博士的名言，直到今天都是多菲酒庄管理的核心理念："优质的葡萄酒诞生在田间，而不是酒窖里"。同其他顶级名庄一样，多菲酒庄把最大的精力和资源放到了葡萄园的培育和管理上。

在升级项目启动之前，葡萄园里有着现存 70 年以上的各种葡萄藤，有意大利本土品种大桑娇维塞，也有梅洛和品丽珠这些大家熟悉的品种。项目团队为了最大限度地保留本地风土，采用了混合选择法 (massal selection)，精心挑选了最优质的葡萄藤进行重新扦插培育。如今，葡萄园里的绝大多数葡萄藤都超过 20 年的历史了。

酿造与陈年

多菲酒庄的白葡萄酒为去皮去梗发酵，红葡萄酒则是带皮去梗发酵。全部采用中性人工酵母启动发酵，保证发酵的稳定性和完成度。

新酒在完成初步发酵以后糖分完全分解成酒精，此时的红葡萄酒还需要经过一次轻柔的压榨环节，将带皮发酵的果皮和酒液进行分离。接下来就进入苹果酸—乳酸发酵 (MLF) 过程，令葡萄酒的口感得到进一步的优化。全部发酵过程中，酒的温度始终控制在 20℃上下。

在陈年的过程中，根据实际情况，一年中还会进行多次的换桶工作。分装在不同容器或橡木桶中的酒液会逐渐产生一些细微的差异，各个桶的蒸发速度也不同，因此需要定期换桶和混合，同时以自然方法去除杂质，用非过滤的方法让酒液变得越来越纯净。

多菲酒庄的装瓶流程非常细致复杂，酒庄会选择晴天、气候干燥、月亮下行的时间，保持酒温在 20℃，从而保证酒的体积后续不会产生剧烈变化。酒塞和酒在装瓶时保持 14 毫米间距，装瓶之后先直立摆放，等酒塞在 3~5 天完全贴合瓶口之后再改为平行摆放，尽量避免因过早横放导致的漏液问题。

在浩瀚的葡萄酒世界里，多菲酒庄不媚俗，不盲从

认认真真做自己的事，安安静静酿自己的酒

纯粹、清澈、天然、真实、自我、执着

把握住生命中最重要的事物，一起品味杯酒人生

关注"多菲酒庄"微信公众号，了解更多资讯

MACCHERONCINI SCAMPI E CECI
HOMEMADE MACCHERONCINI WITH
SCAMPI AND CHICKPEAS

意式手工通心粉配螯虾及鸡心豆

意大利通心粉
细粒小麦粉 500 克
水 300 毫升

For the maccheroncini
500g semolina
300ml water

酱汁
特级初榨橄榄油 100 克
大蒜适量
迷迭香适量
干鹰嘴豆 100 克
鱼汤 200 克
新西兰螯虾 1 公斤
盐适量
白胡椒适量

For the sauce
100g extra virgin olive oil
Garlic, to taste
Rosemary, to taste
100g dried chickpeas
200g fish bouillon
1kg New Zealand scampi
Salt, to taste
White pepper, to taste

将干鹰嘴豆在凉水中浸泡一晚

Soak the chickpeas in cold water overnight

餐盘来自 LEGLE PORCELAIN（法国丽固）地中海系列 Mediterranee
系列。大面积运用的蓝与白，清澈无瑕，诠释着人们对蓝天白云、碧海金沙的无尽渴望。从造型别
致的拱廊得到灵感，形成独特的浑圆造型。

意 大 利 通 心 粉	将小麦粉和水混合，揉成面团。待面团静置松弛 30 分钟后，将其揪成小面团，使用通心粉制作工具将小面团搓成细管状。把做好的通心粉放在撒有小麦粉的砧板上备用。

Prepare the maccher-oncini

Combine the water and the semolina in a large bowl and mix to a firm dough. Rest the dough for 30 minutes and then shape it into small cylinders, pushing a skewer through to finish the shape. Put the maccheroncini on a tray dusted with semolina and leave to rest.

酱 汁	锅中倒入橄榄油，放入大蒜和迷迭香翻炒。待大蒜呈金黄色后，加入鹰嘴豆，翻炒 10 分钟，再加入鱼汤，然后将全部汤料倒入搅拌机中搅拌直至顺滑，静置保温。 将鳌虾洗净去壳，加入少许大蒜一起炒熟，之后切块。 另取锅，将通心粉煮至弹牙。将虾块与鹰嘴豆泥一起翻拌均匀，和通心粉一起装盘，最后淋上少许橄榄油。

Prepare the sauce

In a medium-sized pan, heat the olive oil with the garlic and rosemary. When the garlic is golden, add the chickpeas and cook for ten minutes, add the fish bouillon, and then blend in a food processor until smooth. Keep warm.
Now, clean and peel the scampi. Roast the scampi in a pan with some garlic, and then dice it.
In another pot, cook the maccheroncini until al dente and then sauté with the chickpea cream and the roasted scampi, then serve with the maccheroncini. Finish with olive oil.

烤石斑鱼配新鲜番茄及橄榄

石斑鱼 1 公斤	1kg grouper
樱桃番茄 300 克	300g cherry tomatoes
塔加斯卡橄榄 300 克	300g Taggiasca olives
大蒜 4 瓣	4 garlic cloves
百里香适量	Thyme, to taste
白葡萄酒 1 杯	1 glass white wine
特级初榨橄榄油 100 克	100g extra virgin olive oil
柠檬 1 个	1 lemon

制　　作　　将石斑鱼洗净，去除内脏和鱼鳞，确保鱼头周围清洗干净。
除柠檬外，将所有食材放入烤盘中，用锡箔纸覆盖。放入预热至 200℃的烤箱中，至少烤 25 分钟后取出，将食材放入盘中。
将烤鱼过程中产生的汤汁作为调汁浇在食材上，将柠檬切瓣，点缀在盘中即可。

Prepare
the dish

Clean the grouper, removing the innards and the scales: be sure to clean very well near the head.
Place all the ingredients except the lemon on a baking tray. Cover the tray with foil and bake at
200°C for at least 25 minutes, remove from the oven and place on a plate.
Use the cooking juice to sauce the fish, and serve with lemon wedges.

朗姆巴巴

巴巴面团
高筋面粉 1 公斤
啤酒酵母 50 克
鸡蛋 1.2 公斤
白砂糖 100 克
黄油 400 克
盐 15 克

浸泡糖浆
水 1 千克
白砂糖 250 克
香草荚 1 根
八角茴香 1 个
丁香 10 根
肉桂 1 根
橙皮 1 片
柠檬皮 1 片
黑朗姆酒 250 克

马斯卡彭鲜奶油
马斯卡彭奶酪 100 克
奶油 200 克
卡仕达酱 200 克

For the baba dough
1kg strong bread flour
50g beer yeast
1.2kg whole eggs
100g sugar
400g butter
15g salt

For the soaking syrup
1kg water
250g sugar
1 vanilla pod
1 star anise
10 cloves
1 cinnamon stick
1 piece orange peel
1 piece lemon peel
250g dark rum

For the mascarpone Chantilly
100g mascarpone
200g cream
200g pastry cream

巴巴面团	将高筋面粉、啤酒酵母和一半鸡蛋混合，用搅拌机以中等速度搅拌，在加入剩余鸡蛋之前，形成很好的面筋结构是很重要的。

将高筋面粉、啤酒酵母和一半鸡蛋混合，用搅拌机以中等速度搅拌，在加入剩余鸡蛋之前，形成很好的面筋结构是很重要的。

待面团将第一个鸡蛋完全吸收后，加入白砂糖和剩余鸡蛋。在放入下一个鸡蛋之前，一定要确保面团已完全吸收之前的蛋液。

在还剩最后两个鸡蛋的时候，你要确定面团已经完全呈现发酵的状态，此时把黄油切成小块，与盐混合后一点点地放进去。随后提高搅拌机的速度，让面团有足够的时间吸收黄油，但不要把速度调得太快。将所有的黄油都放入之后，再放入剩下的两个鸡蛋，每次放一个。最终你得到的应该是非常软的面团，类似液体，但在加工时能够形成均匀的团块。

选择你喜欢的大小和形状的巴巴，将它放在温暖的环境里进行定型，定型的时间取决于面团的大小，通常需要 1 小时左右。定型后，你就可以开始烘焙巴巴了，烤箱的温度根据巴巴的大小而变化。对于一个中等大小的巴巴，建议温度为 180℃，烤制 40~50 分钟，最终的成品应该坚硬牢固，呈深棕色。

烤好后，将巴巴取出，放在架子上冷却。

Prepare the baba dough

In the bowl of a stand mixer, combine the strong bread flour, beer yeast, and half of the eggs and start to mix on medium speed. It's important to develop a strong gluten structure before adding the other eggs.

Once the dough starts to develop and the first eggs are completely absorbed, add the sugar and the other eggs one at the time, always making sure the dough absorbs the egg completely before adding the next one.

Once you have two eggs remaining and you are sure the gluten is fully developed, add the salt and the soft butter in small pieces, little by little. Raise the speed of the mixer and make sure the dough has enough time to absorb the butter, but don't make the speed too fast. After all the butter has been incorporated, add the two remaining eggs one at the time. The final result should be a very soft and liquid dough that is able to form a uniform mass when worked.

Form the baba to the size and shape you prefer and place it in a warm spot to prove. The proofing time will vary, depending on the size you choose, but will usually be around one hour. Once proofed, you can proceed to bake the baba. For a medium size baba, I suggest a temperature of 180°C for around 40-50 minutes. The final product should be hard and firm, and have a deep brown colour.

Once baked, remove it from the tin and leave it to cool on a rack.

浸泡糖浆

在锅中将除黑朗姆酒之外的所有食材混合后煮沸，之后关火，再加入黑朗姆酒。随后取出所有香料，趁糖浆还热的时候放入巴巴，将其完全浸泡在糖浆里，放入冰箱冷藏，可以保存一周。

Prepare the soaking syrup

In a large pot, mix all the ingredients except the dark rum and bring to boil. Once boiling, turn off the heat and add the dark rum. Remove all the spices and while the syrup is still hot, add the baba and soak them completely. You can now reserve the baba in the fridge, fully immersed in this syrup, for up to one week.

马斯卡彭 鲜奶油	在搅拌机中混合马斯卡彭奶酪和奶油，开始搅拌，直到达到柔软蓬松的稠度。将搅拌好的奶油与卡仕达酱混合，再搅拌 3 次，注意不要过度搅拌。把混合物放在冰箱里，直到你准备要装盘。

Prepare the mascarpone Chantilly	In the bowl of a stand mixer, combine the mascarpone and the cream and whip until you get a soft and fluffy consistency. Add the whipped cream to the pastry cream in 3 batches, taking care not to over mix it. Reserve in the fridge until you are ready to plate your baba.

装　　盘	从糖浆里取出一个巴巴，挤压出多余的液体。用刀从巴巴中间切开，注意只是切开，而不是切成两半。把切开的巴巴放在盘子里，在巴巴中填满马斯卡彭鲜奶油，再用一些新鲜的红色浆果装饰即可。

Plating	Take one baba from the syrup and squeeze out the excess liquid. Using a knife, cut the baba along the middle to open but don't cut it all the way in half. Place the open baba on a plate and fill the cut with some of the mascarpone Chantilly cream. Garnish the plate with red berries.

八月

Agosto
August

我劝大家八月时尽量不要来意大利，这时的意大利就跟中国春节期间的情形差不多，大部分的商店、企业甚至政府部门都关门了，因为全国都放假啦。意大利人休假无非两种选择：海边度假或山林隐居。从古至今，海边和山林都是避暑的好地方。

Although Italy is beautiful in August, it may be better not to visit at this time of year because the whole country is on holiday and most shops, businesses, and even government offices are closed — it is just like China during Spring Festival! At this time, most Italians travel to either the beach or the mountains. These have always been the best places for holidays.

不过，八月的意大利还是有很多好吃的，几乎所有的餐馆都有海虹供应，这个季节的海虹尤为肥美，怎么烹饪都好吃。天气炎热，人们容易没有胃口，没有什么比鲜凉凉的生牛肉薄片更令人味蕾大开了。我的至爱——鸡油菌在八月时也开始陆续上市了，小时候我还跟小伙伴们一起去山里采过蘑菇呢。此外，这个季节最棒的水果就是桃子了，意大利人对桃子的热爱不亚于中国人，让我欣慰的是，在北京同样可以找到跟意大利一样甜美多汁的桃子，于是在北京我也可以做出家乡夏天特有的桃子甜品啦。

However, there is a lot to eat in Italy in August. The juicy mussels are the best this month, and you can find them in almost all restaurants. Mussels can be cooked in various ways, but they are always tasty. At this time of the year, a chilled beef carpaccio can also rescue appetites stolen away by the summer heat. My favourite chanterelles are also ripe in August. I can still recall when I was a kid, I went to the mountains with my friends to pick mushrooms. August also is the best month for peaches. Italians go crazy for peaches, and love peaches as much as the Chinese do. When I came to Beijing, I was delighted that I could find peaches that were just as sweet and juicy as the ones in Italy. I could make desserts with peaches in Beijing, which are specialties of my hometown.

Menu

小牛肉薄片配芹菜根及帕玛森芝士
Carpaccio di Fassone Parmigiano e Sedano Rapa
Fassone Veal Carpaccio with Celery Root and Parmigiano Reggiano

日本温泉鸡蛋配鸡油菌及洋蓟
Uovo, Chanterelle e Topinambur
Confit Egg, Chanterelles and Artichokes

意式扁面配蒙特圣米歇尔海虹及地中海香草
Linguine alle Cozze
Linguine with Mont Saint-Michel Mussels and Mediterranean Herbs

烤黄狮鱼和时令蔬菜
Ricciola
Seared Amberjack with Seasonal Vegetables

意式传统奶油蛋糕
Zuppa Inglese
Alchermes Liqueur, Chocolate Cream, Vanilla Cream and Strawberry Sorbet

CARPACCIO DI FASSONE PARMIGIANO E SEDANO RAPA
FASSONE VEAL CARPACCIO WITH CELERY ROOT AND PARMIGIANO REGGIANO

小牛肉薄片配芹菜根及帕玛森芝士

新鲜现磨帕玛森芝士 100 克	100g freshly grated Parmigiano Reggiano
小牛里脊 600 克	600g Fassone veal tenderloin
芹菜根 200 克，切片	200g celery roots, sliced
茴香头 200 克，切片	200g fennel bulbs, sliced
第戎芥末酱 100 克	100g Dijon mustard
柠檬汁适量	Lemon juice, to taste
特级初榨橄榄油 100 克	100g extra virgin olive oil
盐适量	Salt, to taste
黑胡椒适量	Black Pepper, to taste
小黄花适量	Yellow flowers, to garnish

帕 玛 森
芝士脆片

在盘子上垫上烘焙纸，放上新鲜现磨帕玛森芝士，放入微波炉中加热 10~15 秒。当芝士变脆时取出（颜色不会发生太大变化），放置在阴凉干燥的地方。

Prepare
the
Parmigiano
Reggiano
crisps

Put the grated Parmigiano Reggiano on a flat plate with some baking paper and microwave it for 10~15 seconds. When it is crispy (the colour does not change too much), set aside in a cool and dry place.

制　　作

将小牛里脊切成圆形薄片，用锤肉器敲成极薄的一层。
将切成片的芹菜根和茴香头用橄榄油、柠檬汁、盐和黑胡椒调味。
将芹菜根和茴香头摆在盘子中间，用牛里脊片覆盖。再在上面刷上少许橄榄油，撒上少许盐。
用帕玛森芝士脆片、小黄花和几滴第戎芥末酱做点缀。

Prepare
the dish

Thinly slice the meat and pound to very thin, round pieces with a kitchen hammer.
Season the celery root and fennel bulbs with olive oil, lemon juice, salt, and black pepper.
Place the vegetables in the centre of the plate and cover with the meat. Brush with olive oil and sprinkle with some salt.
Garnish the plate with Parmigiano Reggiano crisps, yellow flowers, and drops of Dijon mustard.

UOVO, CHANTERELLE E TOPINAMBUR
CONFIT EGG, CHANTERELLES AND
ARTICHOKES

日本温泉鸡蛋配鸡油菌及洋蓟

洋蓟 600 克	600g artichokes
黄油 200 克	200g butter
大蒜 2 瓣	2 garlic cloves
鸡汤 100 毫升	100ml chicken stock
特级初榨橄榄油 100 克	100g extra virgin olive oil
日本温泉鸡蛋（或可生食鸡蛋）6 个	6 Japanese eggs(or any eggs that are safe to eat raw)
新鲜鸡油菌 400 克	400g fresh chanterelles
百里香适量	Thyme, to taste
海盐适量	Sea salt, to taste

洋 蓟 酱

将洋蓟去皮，切成大约 0.5 厘米厚的丁。
在平底锅中放入 1 瓣大蒜和 100 克黄油，再加入洋蓟丁和适量鸡汤。待洋蓟丁煮至软嫩时，取出一部分用于装饰，另一部分搅拌至润滑，保温放置。

Prepare the artichoke cream

Cut the artichokes into 0.5 cm cubes.
In a sauté pan, roast one clove of garlic in half of the butter. Add the artichoke cubes and some chicken stock. When tender, remove some of the cubes for garnish and blend the rest to a smooth puree. Keep warm.

制 作

平底锅中放入橄榄油、剩余黄油、1 瓣大蒜、百里香及鸡油菌煸炒。
将鸡蛋放入 64℃的水中煮至少 25 分钟。
在一个深盘中放入洋蓟酱，再放入用于装饰的洋蓟丁和鸡油菌。
将鸡蛋剥壳后放在蔬菜上，用海盐和特级初榨橄榄油调味即可。

Prepare the dish

In a sauté pan, heat the olive oil and the remaining butter, and roast the remaining garlic clove, the thyme, and the chanterelles.
Cook the eggs in lukewarm water at 64℃ for at least 25 minutes.
In a deep plate, place the warm artichoke cream and add some more cubes of artichoke and the chanterelles for garnish.
Remove the shells from the eggs. Then place the eggs on top of the vegetables and season with sea salt and olive oil.

LINGUINE ALLE COZZE
LINGUINE WITH MONT SAINT-MICHEL
MUSSELS AND MEDITERRANEAN HERBS

意式扁面配蒙特圣米歇尔海虹及地中海香草

特级初榨橄榄油 100 克	100g extra virgin olive oil
大蒜 3 瓣	3 garlic cloves
辣椒 1 个	1 Chilli
百里香适量	Thyme, to taste
海虹 1 公斤	1kg Mont Saint-Michel Mussels
樱桃番茄 300 克	300g cherry tomatoes
意式扁面 400 克	400g linguine Verrigni
欧芹 20 克，切碎	20g parsley, chopped
柠檬汁 1 勺	1 tsp lemon Juice
黄油 40 克	40g butter
盐适量	Salt, to taste
黑胡椒适量	Black pepper, to taste

制　作　平底锅中放入橄榄油、辣椒、大蒜和百里香一同翻炒，再加入洗净的海虹，盖上锅盖，煮至海虹全部张开。
将锅中的汤汁过滤到碗中，备用。再次加热海虹，加入切好的樱桃番茄碎、盐、黑胡椒。
另取锅，将意式扁面放入沸水中煮 8 分钟后捞出。在煮海虹的锅中加入少许过滤好的汤汁，再加入煮好的扁面、欧芹碎、黄油、柠檬汁，再煮几分钟，取出盛盘即可。

Prepare
the dish　Heat the olive oil in a sauté pan with the chilli, garlic, and thyme. Add the cleaned mussels, cover the pan, and wait until the mussels are all open.
Remove the mussels and strain the remaining juice. Bring the juice to boil, adding the chopped cherry tomatoes, salt, and black pepper.
Boil the linguine for 8 minutes in another pot and drain. Then add the filtered juice, the linguine, the chopped parsley, the butter, and the lemon juice into the pot which was used to boil the mussels. Cook for another few minutes and plate the pasta with the mussels on top.

烤黄狮鱼和时令蔬菜

新鲜黄狮鱼 800 克	800g fresh Amberjack
橄榄油 40 克	40g olive oil
大蒜 1 头	1 garlic clove
百里香适量	Thyme, to taste
茴香头 300 克，切块	300g fennel bulbs, julienned
茴香籽 1 咖啡匙	1 tsp fennel seeds
橙子 1 个	1 orange, juiced
混合叶菜 50 克	50g mixed leaves
意大利黑醋 2 汤勺	2 tbsp balsamic vinegar
豌豆 50 克	50g green peas
白芦笋 50 克，切豌豆大的丁	50g white asparagus, pea-sized dice
绿芦笋 50 克，切豌豆大的丁	50g green asparagus, pea-sized dice
芹菜梗 50 克，切豌豆大的丁	50g celery, white part only, pea-sized dice
柠檬汁 20 克	20g lemon Juice
去核绿橄榄 50 克	50g pitted green olives
盐适量	Salt, to taste
白胡椒适量	White pepper, to taste

茴 香 酱	将黄狮鱼清洗干净，把鱼肉切成每片约 80 克的薄片，保留鱼皮，放入冰箱冷藏。 锅中放入 1 勺橄榄油、1 瓣大蒜、少许百里香，放入茴香头块和茴香籽，煮几分钟后挤入橙汁，慢炖 20 分钟，如果汁水少了，就适量加水。 煮好后冷却，将混合物放入搅拌机中搅打，直到呈现奶油质地，备用。
Prepare the fennel purée	Wash and clean the Amberjack. Cut into thin pieces weighing about 80g each, keeping the skin on. Set aside in the fridge. Heat a spoonful of olive oil in a pot and add the garlic and thyme. Add the fennel bulbs and the fennel seeds, and cook for a couple of minutes. Add the orange juice and stew gently for 20 minutes. If it gets too dry, add a little water. Let the fennel cool down and blend to a purée in a food processor. Strain it and keep on one side.
制 作	中火加热不粘锅，在黄狮鱼片上刷上橄榄油，撒上盐和白胡椒，将鱼在锅中带皮煎 30 秒，然后翻至另一面煎，煎好后将锅从火上移开并保持热度。 在大碗里将混合叶菜、盐、白胡椒、橄榄油和意大利黑醋拌匀。 在另一个碗里放入豌豆、白芦笋、绿芦笋、芹菜根、绿橄榄，与柠檬汁、盐和白胡椒充分搅拌。 在盘子里放两勺茴香酱，上面放一片黄狮鱼，再放上一层混合蔬菜，最后放上适量混合叶菜。
Prepare the dish	Place a non-stick pan over a medium heat. Brush the amberjack slices with olive oil, salt, and white pepper. Cook the fish for 30 seconds with skin side down, then flip over and remove from the heat. Keep warm. In large bowl, prepare the mixed leaves with salt, white pepper, olive oil, and balsamic vinegar. In another bowl, toss the diced vegetables with lemon juice, salt, and white pepper. Put two spoons of fennel purée on each plate and top with a slice of Amberjack. Cover with the diced vegetables and top the plate with some mixed leaves.

ZUPPA INGLESE
ALCHERMES LIQUEUR, CHOCOLATE CREAM,
VANILLA CREAM AND STRAWBERRY SORBET

意式传统奶油蛋糕

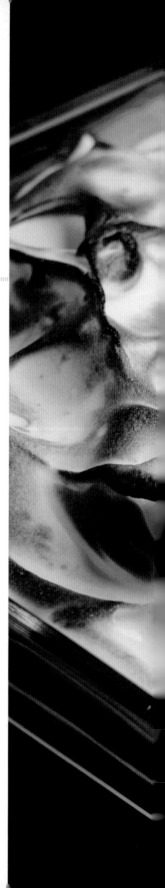

海绵蛋糕
鸡蛋 250 克
白砂糖 175 克
蛋糕粉 150 克
马铃薯淀粉 50 克

For the sponge cake
250g eggs
175g sugar
150g cake flour
50g potato starch

糖浆
水 250 毫升
白砂糖 125 克
利口酒 75 克

For the Alchermes syrup
250ml water
125g sugar
75g Alchermes

香草卡仕达酱
奶油 250 克
牛奶 250 毫升
香草荚 1 根
蛋黄 200 克
白砂糖 125 克
玉米淀粉 30 克

For the vanilla pastry cream
250g cream
250ml milk
1 vanilla pod
200g egg yolks
125g sugar
30g corn starch

巧克力卡仕达酱
牛奶 500 毫升
蛋黄 90 克
白砂糖 125 克
玉米淀粉 20 克
70% 黑巧克力 210克，
切成小块

For the chocolate pastry cream
500ml milk
90g egg yolks
125g sugar
20g corn starch
210g 70% dark chocolate, finely
chopped

意大利蛋白酥　　**For the Italian meringue**
白砂糖 150 克　　150g sugar
水 60 毫升　　60ml water
蛋白 90 克　　90g egg whites

海绵蛋糕　用搅拌机打发鸡蛋和白砂糖，将蛋糕粉和马铃薯淀粉混合过筛，一旦鸡蛋打发至变白且质地蓬松，分三次加入蛋糕粉混合物。将制成的面糊倒入蛋糕模具或烤盘中，在烤箱中以 180℃烘烤 15~20 分钟，具体时间取决于烤箱性能。烤好后，将海绵蛋糕从模具中取出，放在架子上冷却。如果您当天制作蛋糕，就用保鲜膜包裹好蛋糕坯，放在室温下保存。如果您暂时不用，则可以在冰箱中最多冷藏放置一个月 。

Prepare the sponge cake　In the bowl of a stand mixer, whip together the eggs and the sugar. Sift together the cake flour and potato starch. Once the eggs are pale and fluffy, add the cake flour mixture in three batches. Pour the batter into a cake tin or a tray and bake in the oven at 180℃ for around 15~20 minutes. Once baked, remove the sponge cake from the mould and let it cool down on a rack before wrapping in plastic film. Keep at room temperature if you are assembling the cake the same day, or in the freezer for up to one month.

| 糖　浆 | 锅中混合水和白砂糖，加热至白砂糖完全溶解，关火，冷却后加入利口酒。将糖浆放入冰箱冷藏，直到您准备使用时。 |

Prepare
the
Alchermes
syrup

In a pot, mix water and sugar. Warm until the sugar completely dissolves. Leave to cool slightly and then add the Alchermes. Reserve the syrup in the fridge until ready to use.

| 香　草
卡仕达酱 | 锅中混合奶油、牛奶、香草荚，煮沸。将蛋黄、白砂糖和玉米淀粉混合。
待牛奶混合物沸腾后，分三次倒在蛋黄混合物中并持续搅拌，让蛋黄混合物温度慢慢升高。 然后将全部混合物倒回锅中，用中低火煮沸，同时不断搅拌。
一旦混合物开始沸腾，随即关火，继续搅拌几分钟，此时锅中混合物呈浓稠奶油状，光滑无结块。将卡仕达酱转移到一个扁平的容器中，用保鲜膜覆盖，避免凝结，放入冰箱冷藏，直到准备使用时。 |

Prepare
the vanilla
pastry
cream

Combine the cream, milk, and the vanilla in a pot and bring to the boil. Mix the egg yolks with the sugar and corn starch.
Once the milk mixture is boiling, pour one third of it over the yolk mixture and whisk to raise the temperature of it gently, followed by another third, and then the final third.
Pour this mixture back into the pot and bring to a simmer over a low heat, stirring constantly.
Once you see the mixture start to simmer, turn off the heat and keep mixing for few more minutes.
At this point you should have a thick but creamy mixture that is smooth and without lumps.
Transfer the pastry cream to a flat container and cover it with plastic film to avoid coagulation.
Reserve the cream in the fridge until you are ready to use it.

| 巧 克 力
卡仕达酱 | 与香草卡仕达酱制作步骤相同，其中不同的是，在卡仕达酱做好后，迅速将黑巧克力块放入卡仕达酱中搅拌，直到黑巧克力完全融化。 将卡仕达酱转移到一个扁平的容器中，用保鲜膜覆盖，避免凝结，放入冰箱冷藏，直到准备使用时。 |

Prepare
the
chocolate
pastry
cream

Follow the same steps for the vanilla pastry cream. Then once the cream is ready, mix in the dark chocolate while the cream is still hot. Transfer the cream to a flat container and cover with plastic film to avoid coagulation. Keep the cream in the fridge until you are ready to use it.

意 大 利
蛋 白 酥

将蛋白倒入搅拌机中，慢慢搅打至起泡。 在小锅中将白砂糖和水煮至 118℃，一旦糖浆达到温度，将搅拌机的速度提高到最大，这时平稳地将热糖浆倒入搅拌机中，搅拌同时运行。 搅拌几分钟后停止，静置到室温，然后将其用保鲜膜覆盖，直到开始装饰甜点。

Prepare
the Italian
meringue

Place the egg whites in a bowl of a stand mixer and start to slowly whip them until they froth up. Cook the sugar and water to 118°C. Once the syrup heats up, raise the mixer's speed to maximum and slowly pour the hot syrup into the bowl of the mixer. Let the machine run for a few minutes, and cool the meringue to room temperature. Reserve the meringue covered with plastic film until ready to decorate your dessert.

组　　装

选一个漂亮的高边陶瓷碗或玻璃碗，这样一旦甜点组装完成，您就可以看到分层。
首先将一些香草卡仕达酱涂抹在容器底部，然后铺一层海绵蛋糕，将其切成所需形状，厚度约为 0.5 厘米。用刷子将糖浆涂抹整个海绵蛋糕，然后在上面用勺子或小抹刀均匀涂抹一层巧克力卡仕达酱。再放一层海绵蛋糕，同样用糖浆涂抹，在上面再涂上一些香草卡仕达酱，并重复相同的步骤，之后再涂一层巧克力卡仕达酱。如果一切正常，您应该得到 4 层卡仕达酱，即 2 层香草的和 2 层巧克力的，以及 3 片海绵蛋糕。
在装饰甜点顶部时，撒上一些意大利蛋白酥，并用勺子背面制作出一些尖刺。最后用一把小喷枪轻轻烧焦蛋白酥，效果极佳。

Assembly

The day you are ready to assemble the dessert, choose a nice ceramic dish with tall edges or a glass dish so you can see the layers once the dessert is finished.
Start by spreading some of the vanilla pastry cream on the bottom of the container, followed by a layer of sponge cake about 0.5 cm thick in the shape you need. Brush the sponge with the Alchermes syrup, and spread over a layer of chocolate cream. Even out the cream with a spatula. Add another layer of sponge cake and soak it with the syrup as before. This time, spread some of the vanilla cream over it and repeat the same process again for another layer of chocolate cream. You should end up with four layers of cream, two vanilla and two chocolate, and three layers of sponge.
To decorate the top of the dessert, spread the top layer with the Italian meringue and create spikes with a spoon. Lightly brown the meringue with a blow torch.

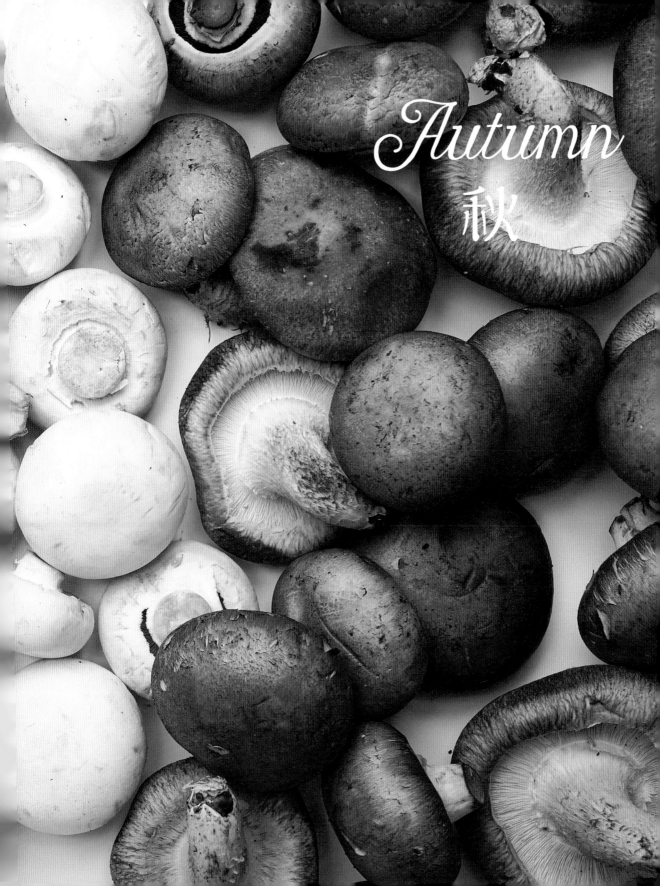

Autumn

秋

九月

秋天来了，意大利的风景也随着植被颜色的多彩而变幻得无比绚丽。人们纷纷回到工作和学习的岗位上，恢复了日常的平静。

九月的无花果异常甜美，我喜欢用无花果配上 24 个月熟成的意大利黑猪火腿（culatello），咸甜相宜，肥美多汁，简简单单却堪称完美。此外，妈妈院子里留下的最后一季的茄子，攒足了一个夏天的浓郁味道，与番茄和水牛奶酪是绝配。不过不用太过留恋，令人兴奋的是，新一季的蘑菇纷纷上市啦！其中我最爱的便是牛肝菌，它是世界上最鲜美的蘑菇之一，也是在意大利仅次于松露的山珍，因肉质肥厚，色如牛肝而得名。牛肝菌可烩饭、可熬制意面酱，也可直接香煎佐餐，怎么做都好吃。

Italy is full of colourful vegetables as autum approaches. People return from holiday to their daily routines, and everything goes on as usual. Fresh figs are at their plumpest and sweetest at this time of year, the perfect partner of savoury culatello slices. Simpe but perfect. My mom always keeps the eggplants of the last summer in our yard. Eggplant is a perfect match for tomatoes and buffalo mozzarella with its tangy flavors of summer. However, you don't need to linger over eggplants too much, because September is also mushroom season. Every year, I get excited for the arrival of porcini, the king of the mushrooms, second only to black truffles in Italy. Whether they are added to a risotto, cooked into a puree to serve with pasta, or simply pan-fried, they are always delicious.

此外，九月还是意大利各地葡萄园丰收的季节，在很多葡萄酒产地的村庄里，人们用聚会、音乐和美食派对来召唤又一个"伟大年份"的到来，庆祝一年的辛勤劳作。

September is "Vendemmia", the harvest time for vineyards across Italy. In many villages, people call out another "great year" and celebrate a year of hard work with gatherings, music, and delicious treats.

Menu

24 个月熟成意大利 Culatello 黑猪火腿及新鲜无花果酱
Culatello e Fichi
Culatello with Fig Compote

经典茄子塔配番茄及水牛芝士
Melanzane alla Parmigiana
Eggplant with Tomato and Buffalo Mozzarella

黄金意面配诺曼底蓝龙虾及樱桃番茄
Spaghetti all'Astice
Spaghetti with Normandy Lobster and Cherry Tomatoes

香烤澳洲牛柳及新鲜牛肝菌
Fileto di manzo e Porcini
Roast Australian Beef Tenderloin with Porcini Mushrooms

卡布里蛋糕
Caprese
Caprese Cake

24个月熟成意大利 Culatello 黑猪火腿及新鲜无花果酱

新鲜无花果 200 克	200g fresh figs
肉桂棒 1 根	1 cinnamon stick
八角茴香适量	Star anise, to taste
香草荚 1 根	1 vanilla pod
葡萄糖 80 克	80g glucose
果胶 20 克	20g pectin
波特酒 400 克	400g Port wine
水 1 升	1L water
黑猪火腿 400 克	400g Culatello
智利无花果 400 克	400g Chilean figs

制　作　将新鲜无花果洗净，切成四瓣后放入锅中，再放入所有香辛料、葡萄糖、果胶、波特酒和 1 升水，保持在 80℃熬煮一晚，即成无花果酱。
盘中放上切成片的黑猪火腿，摆上切成四瓣的智利无花果，然后淋上无花果酱即可。

Prepare the dish

Wash and quarter the fresh figs. Place them in a large pot with the spices, glucose, pectin, Port wine, and 1L water. Cook at 80°C overnight. This is the fig compote.
Cover a plate with thin slices of Culatello. Place the quartered Chilean figs on top, then drizzle with fig compote.

MELANZANE ALLA PARMIGIANA
EGGPLANT WITH TOMATO AND BUFFALO
MOZZARELLA

经典茄子塔配番茄及水牛芝士

圆茄子 2 公斤	2kg round eggplants
面粉 200 克	200g flour
葵花籽油 3 升	3L sunflower oil
黄油 50 克	50g butter
水牛芝士 400 克，切块	400g buffalo mozzarella, cubed
罗勒叶 50 克	50g basil leaves
新鲜现磨帕玛森芝士 250 克	250g freshly grated Parmigiano Reggiano
番茄酱 500 克	500g tomato sauce
盐适量	Salt, to taste
黑胡椒适量	Black Pepper, to taste

制　作　　将茄子洗净，切成 0.5 厘米厚的片。把茄子片放在一个滤盆里，撒上盐，放入冰箱冷藏一晚。
将茄子从冰箱中取出，倒掉多余的水。
将茄子用面粉裹好，放入 180℃的热油中炸制，之后放在厨房用纸上静置。
将黄油均匀涂抹在烤盘内，铺上茄子，在茄子上放上水牛芝士块、罗勒叶和帕玛森芝士。最后一层用黄油和帕玛森芝士一同覆盖。放入烤箱中，180℃烤 30 分钟。
与此同时，把番茄酱、黄油和少量帕玛森芝士一同在锅中熬煮。
将烤好的茄子切成方块，淋上少许熬好的番茄酱，点缀上新鲜的罗勒叶即可。

Prepare
the dish

Wash and cut the eggplants into 0.5 cm-thick slices. Place the eggplant slices in a colander and sprinkle with plenty of salt. Rest in the fridge overnight.

Remove the slices from the fridge and discard any water that has accumulated.

Dust the eggplant slices with flour and deep-fry in hot oil at 180°C. Rest on kitchen paper.

Coat the inside of an oven tray with butter. Layer the eggplant slices in the tray with the buffalo mozzarella cubes, basil leaves, and Parmigiano Reggiano. Cover the last layer with butter and Parmigiano Reggiano. Bake for 30 minutes at 180°C.

Simmer the tomato sauce, butter and a small amount of Parmigiano Reggiano together.

Cut the baked eggplant into squares and serve topped with the tomato sauce, garnished with fresh basil leaves.

SPAGHETTI ALL'ASTICE
SPAGHETTI WITH NORMANDY LOBSTER
AND CHERRY TOMATOES

黄 金 意 面 配 诺 曼 底 蓝 龙 虾 及 樱 桃 番 茄

诺曼底蓝龙虾 1 只	1 Normandy blue lobster
特级初榨橄榄油 100 克	100g extra virgin olive oil
百里香适量	Thyme, to taste
大蒜 2 瓣	2 garlic cloves
白兰地 30 毫升	30ml brandy
樱桃番茄 200 克	200g cherry tomatoes
龙虾高汤适量	Lobster stock, to taste
意大利面 500 克	500g spaghetti
干乌鱼籽 20 克	20g bottarga

制　　作　　将龙虾放在沸水里煮 3 分钟，然后放入冰水里冷却。去掉龙虾壳和内脏，再把虾钳和虾尾中的肉取出。
平底锅中加入橄榄油、大蒜和百里香，当大蒜煎至金黄时，将百里香取出，放入龙虾肉，翻炒 30 秒。淋上少许白兰地，然后将龙虾肉取出备用。锅中加入樱桃番茄和适量龙虾高汤。
另取锅，将意大利面放入水中煮 9 分钟。将意面捞出，放入刚炒好的酱汁锅中，再放入龙虾肉拌匀。摆盘时淋上少许橄榄油，点缀少许干乌鱼籽即可。

Prepare
the dish
Boil the lobster for 3 minutes and then place in a bath of iced water. Remove the shell and the intestines, and then set aside the claw and tail meat.
Heat the olive oil in a sauté pan with the thyme and the garlic. When the garlic is golden, remove the herbs and cook the lobster meat for 30 seconds. Sprinkle with brandy and remove the lobster from the pan and set aside. Now, add the cherry tomatoes and some lobster stock.
Boil the spaghetti for 9 minutes in another pot, and then add the spaghetti to the sauce pan. Add the lobster meat and mix well. Finish the pasta with some olive oil and some freshly grated bottarga.

FILETO DI MANZO E PORCINI
ROAST AUSTRALIAN BEEF TENDERLOIN
WITH PORCINI MUSHROOMS

香烤澳洲牛柳及新鲜牛肝菌

澳洲和牛里脊 800 克	800g wagyu beef tenderloin
新鲜牛肝菌 400 克	400g fresh porcini
黄油 200 克	200g butter
大蒜 2 瓣	2 garlic cloves
土豆 200 克	200g potatoes
牛奶 50 毫升	50ml milk
迷迭香 20 克	20g rosemary
牛肉酱汁 100 克	100g beef sauce
橄榄油适量	Olive oil, to taste
盐适量	Salt, to taste
现磨黑胡椒适量	Freshly ground black pepper, to taste

将和牛里脊切成 4 份，每份 200 克。用盐、黑胡椒调味后，放入冰箱冷藏。

将牛肝菌切成 0.5 厘米厚的片，放入平底锅中，加入少许黄油和大蒜，煎至上色，然后从锅中取出，备用。

制作土豆泥。先将土豆放在烤盘中，加适量盐和少量水，放入烤箱 200℃烤制，当土豆变软时即可取出，去皮。用捣碎器将土豆捣碎，然后放入一口小锅中，加入温热的牛奶和黄油，搅拌至质地均匀、顺滑。

把余下的黄油、橄榄油和迷迭香一同放入平底锅中炒出香气，再把和牛里脊放入锅中，煎至您喜欢的熟度，然后在和牛里脊上涂抹黄油。

在盘中放上一勺土豆泥，再放上牛肝菌与和牛里脊，然后淋上牛肉酱汁，撒上适量盐。

Prepare the dish

Cut the wagyu beef tenderloin into four 200g pieces. Season well with the salt and the black pepper, and keep in the fridge.

Cut the porcini into 0.5 cm-thick slices. Roast the porcini in a sauté pan with a little butter and some garlic until they take on a nice colour and then remove from the pan and set aside.

Prepare the mashed potato. Cook the potatoes in the oven at 200°C with some salt and a bit of water in an oven tray. When the potatoes are soft, remove from the oven and peel them. Pass through a masher. Place the mash in a small pot with the warm milk and the butter and mix until nice and fluffy.

Heat the remaining butter, the olive oil, and the rosemary in a sauté pan and roast the wagyu beef tenderloin until cooked to your liking, using a spoon to baste with the butter.

Place a spoon of mashed potato on a plate along with the roasted porcini and the tenderloin, then finish the dish with beef sauce and salt.

CAPRESE

CAPRESE CAKE

卡布里蛋糕

巧克力蛋糕
黄油 175 克
70% 黑巧克力 245 克
蛋黄 105 克
杏仁粉 175 克
榛子粉 87 克
泡打粉 14 克
白砂糖（A）115 克
蛋白 126 克
白砂糖（B）115 克

For the chocolate cake
175g butter
245g 70% dark chocolate
105g egg yolks
175g almond meal
87g hazelnut meal
14g baking powder
115g sugar (A)
126g egg whites
115g sugar (B)

巧克力酱
牛奶 500 毫升
蛋黄 90 克
白砂糖 125 克
玉米淀粉 20 克
70% 黑巧克力 210 克

For the chocolate pastry cream
500ml milk
90g egg yolks
125g sugar
20g corn starch
210g 70% dark chocolate

装饰
杏仁或浆果
糖粉适量

For plating
Nuts or berries
Icing sugar, to taste

| 巧 克 力 蛋 糕 | 将黄油和黑巧克力放在碗中，放入微波炉加热融化，之后加入蛋黄。
在另一个碗中将杏仁粉、榛子粉、泡打粉和白砂糖（A）混合，再加入巧克力混合物，充分搅拌均匀。
用搅拌机将蛋白和白砂糖（B）充分打发，做成蛋白霜。
把蛋白霜拌入面糊中。在小蛋糕模具里抹少许黄油，把蛋糕糊倒进去。烤箱预热至170℃，将模具放入烤箱中烤20~25分钟，具体时间取决于你的烤箱。蛋糕烤好后，在室温下静置，直到你准备把它们装盘。 |

Prepare the chocolate cake

Melt the butter and the dark chocolate together in a microwave, then mix in the egg yolks.

In another bowl, mix together the almond meal, the hazelnut meal, the baking powder, and the sugar (A). Add the chocolate mixture and mix thoroughly.

Using a stand mixer, whisk the egg white and sugar (B) to make a meringue.

Fold the meringue into the batter. Oil some small cake tins with butter and add the cake batter. Preheat the oven to 170°C and bake the cakes for 20~25 minutes, depending on your oven. Once baked, reserve the cakes at room temperature until you are ready to plate.

巧克力酱

将牛奶倒入锅中煮沸。另取一个盆，将蛋黄、白砂糖和玉米淀粉混合均匀。

将1/3煮沸后的牛奶倒入蛋黄混合物中，轻轻搅拌，待温度升高后再倒入1/3牛奶，一直搅拌，最后再倒入剩余的1/3。把混合物倒回锅里，用小火慢煮，同时不断搅拌。

当你看到混合物开始沸腾时就可以关火了，加入切碎的黑巧克力，继续搅拌，直到黑巧克力全部融化并混合在一起，这时你会得到一锅黏稠的、呈奶油状的膏状混合物，质地光滑，没有结块。

将膏体转移到一个扁平的容器中，表面用保鲜膜覆盖，避免其凝结。放入冰箱冷藏。

Prepare the chocolate pastry cream

Put the milk in a pot and bring to boil. In a separate bowl, mix the egg yolks with the sugar and the corn starch.

Once the milk is boiling, pour one third of it over the yolk mixture and whisk in order to raise the temperature of it gently. Add the remaining two thirds one by one, being sure to keep whisking gently. Pour the mixture back into the pot and bring to a simmer over a gentle heat while stirring constantly.

Once the mixture starts to boil, turn off the heat and add the roughly chopped dark chocolate. Mix for a few more seconds until melted and combined. At this point, you should have a thick and creamy mixture with no lumps.

Transfer the cream to a flat container and cover the surface with plastic film to avoid condensation. Reserve the cream in the fridge until you are ready to use it.

装　饰　　在盘子的一侧摆放一块巧克力蛋糕，在表面挤一层螺旋状的巧克力酱，然后在旁边放一勺半打发好的奶油，用你喜欢的食材装饰盘子，比如杏仁、榛子或红色浆果，最后撒上糖粉。

Plating　　Place one caprese chocolate cake on the side of the plate and pipe a spiral of chocolate pastry cream on the top. Spoon a large quenelle of semi-whipped cream on the side, and decorate the plate to your liking using almonds, hazelnuts and red berries. Finally, sprinkle with the icing sugar.

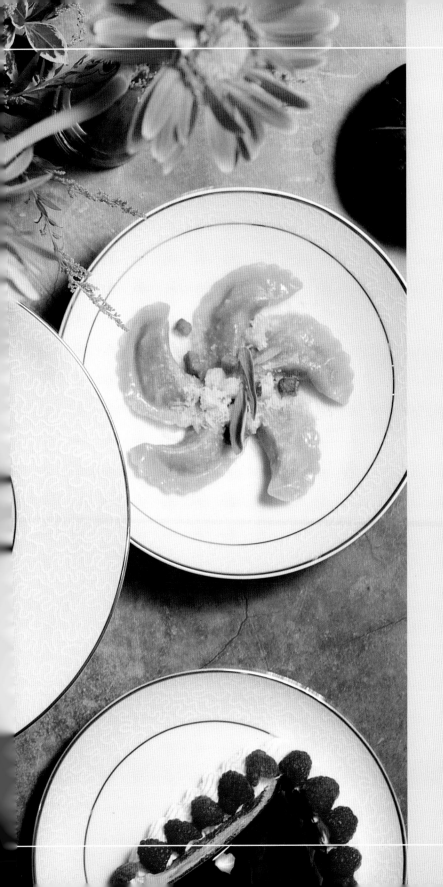

十月

Ottobre
October

金秋十月，好吃的核桃上市
了，我喜欢用核桃佐奶酪吃，
满口留香。意大利的十月，
甜菜根、西兰花和胡萝卜都
正当时。不过，秋天我最喜
欢的还是南瓜，十月的南瓜
拥有非常漂亮的橙色外衣，
如此多姿，如此甜美。我特
别喜欢用南瓜来烹饪时令菜。
记得小时候，妈妈会熬一锅
暖暖的南瓜汤，再撒上一把
杏仁，香气四溢。

在我的家乡贝加莫，因为地
处山区，秋天时常常刮起大
风，预示着冬天的脚步近了，
所以暖汤和饺子又回到了餐
桌上。我的祖母会在这个季
节亲自制作贝加莫特色饺子，
暖暖地吃下去，驱赶寒意，
暖身暖心。

October is walnut season. I like to serve walnuts with good cheese because their flavours pair well in the mouth. In October, beetroot, broccoli and carrots are all in season, but my favourite ingredient this month is pumpkin. At this time of year, pumpkins have a bright orange colour and a versatile set favour, making them perfect to cook with. I remember when I was growing up in Bergamo, my mother would make a pot of pumpkin soup with a handful of almonds, and the sweet fragrance would fill the whole house.

My hometown of Bergamo is located in the mountains, and it can be windy and cold in October, signalling that winter is on the way, so warming dishes like soup or homemade ravioli are back on the menu. My grandmother would make Bergamo-style ravioli at this time which can drive away the cold and warm your body.

如今在意大利，人们也喜欢庆祝万圣节，雕刻南瓜、鬼脸面具和庆祝派对随处可见，非常有趣。

In modern Italy, people also love to celebrate Halloween. Carved pumpkins, masks, and parties can be found all over the cities, and the atmosphere is a lot of fun.

我的祖母 Letizia 是一位了不起的女性，她做的 Casoncelli（意大利饺子的一种）是最好吃的。Casoncelli 是贝加莫传统面食，形状非常接近中国的饺子，里面包裹着肉馅、奶酪、葡萄干等食材，煮熟以后，通常还会在上面淋上浓郁的奶油、培根和鼠尾草等香料，这道菜谱的历史可以追溯到 17 世纪。这道菜也是每年十月我最美好的味蕾记忆。

My grandmother Letizia was an incredible woman, and the casoncelli (a kind of Italian dumpling) she prepared were the best.Casoncelli is a traditional pasta of Bergamo. Its shape is very similar to that of Chinese dumplings. It is wrapped with meat, cheese, raisins and other ingredients. After cooking, it is usually topped with rich cream, bacon, sage and other spices. The history of this recipe can be traced back to the 17th century. This dish is also my best taste buds' memory every October.

Menu

嫩菠菜沙拉配核桃及帕玛森芝士
Spinaci Noci e Parmigiano
Spinach, Walnut and Parmigiano Reggiano Salad

南瓜汤配香烤杏仁碎
Zuppa di Zucca
Pumpkin Soup with Roasted Almonds

意式手工贝加莫祖母饺子
Casoncelli alla Bergamasca
Homemade Ravioli, Bergamo Style

慢炖小羊腿配土耳其杏干、土豆及卡拉玛塔橄榄
Agnello brasato
Braised Lamb Leg with Apricots, Potatoes and Olives

杏仁舒芙蕾
Almond Soufflé

SPINACI NOCI E PARMIGIANO
SPINACH, WALNUT AND
PARMIGIANO REGGIANO SALAD

嫩菠菜沙拉配核桃及帕玛森芝士

新鲜菠菜 300 克	300g fresh spinach
帕玛森芝士 100 克	100g Parmigiano Reggiano
特级压榨橄榄油 80 克	80g extra virgin olive oil
20 年陈年黑醋 60 克	60g 20-year aged Balsamic vinegar
核桃 200 克	200g walnuts
盐适量	Salt, to taste
新鲜现磨黑胡椒适量	Freshly ground black pepper, to taste
小黄花适量	Yellow flowers, to garnish

制　作　将菠菜洗净，摘出品质好的嫩叶子。用切片器把帕玛森芝士切成片。
在碗里将菠菜和帕玛森芝士用橄榄油、盐和黑胡椒搅拌均匀，淋上黑醋。
将拌好的沙拉放入盘中，点缀上核桃、小黄花，再点上几滴黑醋。

Prepare
the dish

Wash the spinach well and select the nice, young leaves. Shave the Parmigiano Reggiano on a mandolin.
In a bowl, toss the spinach and Parmigiano Reggiano with the olive oil, the salt and the black pepper. Add the balsamic vinegar at the last minute.
Put the salad on a plate and garnish with some walnuts, yellow flowers, and drops of balsamic vinegar.

南瓜汤配香烤杏仁碎

南瓜 600 克	600g pumpkins
特级初榨橄榄油 60 克	60g extra virgin olive oil
芹菜 200 克	200g celery
胡萝卜 100 克	100g carrots
蔬菜高汤 200 毫升	200ml vegetable stock
黄油 100 克	100g butter
杏仁片 60 克	60g sliced almonds
盐适量	Salt, to taste
新鲜现磨黑胡椒适量	Freshly ground black pepper, to taste

制作过程

将南瓜切成小块，去除南瓜籽和南瓜皮。放入烤箱 200℃烤 30 分钟。

将芹菜和胡萝卜切成小丁。

在汤锅中加入橄榄油，放入胡萝卜丁和芹菜丁翻炒几分钟，然后加入南瓜块。

加入盐、黑胡椒和蔬菜高汤熬煮，直至汤汁浓缩至三分之一。然后加入黄油，混合熬煮 10 分钟。

将汤汁用细筛过滤，将杏仁片点缀在汤上。

Prapare
the dish

Cut the pumpkins into small pieces and remove the seeds and skin. Bake the pumpkins in the oven for 30 minutes at 200°C.

Cut the celery and carrots into small dice.

Heat the olive oil in a soup pot, sauté the carrots and celery for a few minutes, then add the pumpkins.

Add the salt, the black pepper, and the vegetable stock and reduce the soup to 1/3. Now, add the butter and blend very well and boil for 10 minutes.

Strain the soup through a fine sieve and serve garnished with sliced almond.

CASONCELLI ALLA BERGAMASCA

HOMEMADE RAVIOLI, BERGAMO STYLE

意式手工贝加莫祖母饺子

小牛肩颈肉 300 克	300g veal shoulder
猪颈肉 300 克	300g pork neck
意式杏仁饼 50 克，切碎	50g amaretti biscuits, crumbled
白胡椒适量	White pepper, to taste
细磨肉桂粉适量	Ground nutmeg, to taste
梨 100 克	100g pears
帕玛森芝士 300g	300g Parmigiano Reggiano
马斯卡彭奶酪 80 克	80g mascarpone
新鲜意面面团 500 克	500g fresh pasta dough
黄油 200 克	200g butter
意式咸猪肉丁 200 克	200g cured pancetta
新鲜鼠尾草 30 克	30g fresh sage
盐适量	Salt, to taste
油适量	Oil, to taste

馅　料　将小牛肩颈肉和猪颈肉分别切成 2 厘米厚的丁，用盐和白胡椒调味。在平底锅内加入适量油，将肉丁炒至金黄。

将炒好的肉丁剁碎，把意式杏仁饼、所有香辛料、去皮切成丁的梨，一同放入大碗里。加入帕玛森芝士和马斯卡彭奶酪，混合均匀，直至馅料的质地变得稳固。

Prepare the stuffing

Cut the veal shoulder and pork neck into 2cm cubes and season with the salt and the white pepper. Heat some oil in a sauté pan and brown the meat well.

Cut the meat into a fine brunoise and place in a bowl with the crumbled amaretti, the spices, and the peeled pears cut into small cubes. Add some Parmigiano Reggiano and the mascarpone and mix well until the texture of the filling is firm.

饺　子　将新鲜意面面团擀成很薄的面皮，加入馅料后对折成半圆形状。

将黄油、适量新鲜鼠尾草和肉丁一同放入锅中加热，直至黄油呈泡沫状态。

另取锅，将饺子煮熟，盛出后摆盘，浇上熬好的黄油肉丁，撒上帕玛森芝士和剩余的鼠尾草。

Prepare the ravioli

Roll the fresh pasta dough out very thinly and prepare half moon-shaped casoncelli stuffed with the filling.

Heat the butter, the fresh sage, and the pancetta cubes in a small pan until the butter is foamy.

Cook the casoncelli, then plate up and top with the butter, the Parmigiano Reggiano, and the remaining sage.

AGNELLO BRASATO
BRAISED LAMB LEG WITH
APRICOTS, POTATOES AND OLIVES

慢炖小羊腿配土耳其杏干、
土豆及卡拉玛塔橄榄

小羊腿 8 只	8 lamb shanks
面粉适量	Flour, as required
特级初榨橄榄油 100 克	100g extra virgin olive oil
洋葱 50 克，切片	50g onions, sliced
芹菜 200 克，切丁	200g celery, diced
土豆 500 克	500g potatoes
胡萝卜 200 克	200g carrots
鸡汤 500 毫升	500ml chicken stock
土耳其杏干 200 克	200g dried apricots
黄油 100 克	100g butter
迷迭香适量	Rosemary, to taste

制　作　锅中加热橄榄油，将面粉裹在小羊腿上，煎至两面焦黄。
另取一口锅，加入洋葱、迷迭香和芹菜丁，再加入切成 2 厘米厚的胡萝卜丁和土豆丁，放入土耳其杏干。最后放入小羊腿，倒入鸡汤熬煮至少 2 个小时。
当羊肉变软嫩时，从汤中取出，静置保温。
继续熬煮汤汁直至形成奶油般润滑的质地，将汤汁浇在小羊腿肉上，用蔬菜丁和土耳其杏干装饰即可。

Prepare
the dish　　Heat the olive oil in a large casserole and dust the lamb shanks with flour, browning throughly on all sides.
In another pan, place the sliced onions, the rosemary, and the celery. Add the carrots and the potatoes cut into 2cm cubes, and the dried apricots. Add the lamb shanks to the vegetables, top with chicken stock, and cook slowly for at least 2 hours.
When the lamb is nice and tender, remove from the sauce and keep warm.
Reduce the sauce and when it has a creamy consistency, serve on top of the lamb shanks with some vegetable cubes and apricots as garnish.

杏仁舒芙蕾

舒芙蕾面糊
杏仁奶 200 毫升
奶油 50 克
杏仁酱 60 克
玉米淀粉 20 克
蛋黄 30 克
杏仁利口酒 10 毫升

For the soufflé base
200ml almond milk
50g cream
60g almond paste
20g corn starch
30g egg yolks
10ml Disaronno

蛋白霜
蛋白 150 克
白砂糖 60 克

For the meringue
150g egg whites
60g sugar

组装
黄油适量
白砂糖适量

Assembly
Butter, to taste
Sugar, to taste

舒 芙 蕾 面　　糊	将杏仁奶、奶油、杏仁酱和玉米淀粉放入锅中煮沸，直至完全煮熟。 关火，加入蛋黄和杏仁利口酒。
Prepare the soufflé base	In a pot, boil the milk, cream, almond paste, and corn starch until fully cooked. Turn off the heat. Add the egg yolks and the Disaronno.
蛋 白 霜	将蛋白和白砂糖混合搅拌至中等打发状态。
Prepare the meringue	Whip the egg whites and sugar to medium-hard peaks.
组 装 和 烘　　焙	挑选一个烤碗，在碗壁内涂上足够多的黄油，并在上面撒上白砂糖，这样可以防止面糊粘在边缘上，以便它在烤箱里发酵。把烤碗放在一边或放在冰箱里冷藏，直到你准备好把舒芙蕾面糊倒进去。 将保持温热状态的蛋白霜倒入舒芙蕾面糊中，小心地翻拌三次，注意不要搅拌过度，否则会失去大量的气泡，致使体积减小。把混合物倒入烤碗中，用刮刀将表面整理平整，用拇指在烤碗内侧的边缘划出一条干净的线。 将烤碗放入 230℃的静态烤箱中烘焙几分钟，时间长短取决于烤碗的大小，当舒芙蕾停止膨胀并且不再摇晃时，就已经烤好了。尽快上桌食用，否则它会很快塌陷下去。
Assembly and baking	Take a large ramekin and butter the inside generously and dust it with sugar. This will prevent the soufflé from sticking to the sides and will allow it to rise in the oven. Set the ramekin aside or in the fridge until you are ready to pour in the soufflé. Fold the meringue into the warm base carefully in three batches. Be careful to not over mix, or you will burst the bubbles. Pipe the mixture into the buttered ramekins, flatten the top with an offset spatula, and run your thumb around the inside edge of the ramekin to create a clean line. Bake the soufflé in a static oven at 230°C for few minutes, depending on the size of your ramekin. You know the soufflé is done baking when it stops rising from the cup and is not wobbling anymore. Serve the soufflé right away or it will deflate very quickly.

十一月

Novembre
November

十一月的意大利，寒冷而忙碌，人们努力工作，为圣诞节和新年做准备。

栗子是这个时节的宠儿，无论做汤、烩饭、做甜品，还是烤栗子做零食，怎么做都好吃。值得一提的是，在意大利北部我的家乡贝加莫有意大利最出名的玉米糊（polenta），人们的餐桌上顿顿少不了它。polenta跟中国东北的"棒子面"类似，相比之下，意大利的玉米糊颜色更偏金黄，味道微咸。同样，它也有很多种烹饪方法，或煎或用黄油烤制玉米饼，或熬制成玉米糊，搭配奶酪或奶油蘑菇酱，还可同肉汤一起吃，能够充分地在冬季为身体积攒热量，在寒冬里为我们带来温暖和愉悦，它也是十一月餐桌上的最佳伴侣。

November in Italy may be chilly but we keep warm rushing around making initial preparations for Christmas! Chestnuts are a favourite at this time of year, and they are delicious served in soups, risottos, and desserts, or simply roasted as a snack. Bergamo is also famous for Polenta, a wonderful smooth corn purée that is similar to "cornmeal" from northeastern China. In contrast, however, Polenta is more golden in colour and slightly saltier. As a perfect side dish on the table in November, it can also be served in a variety of ways, such as fried, baked with butter boiled into a paste, or with cheese, creamy mushrooms and broth. It helps warm our bodies and bring us the joy in cold winter.

十一月也是意大利比较寒冷的月份，我的家乡 Cisano Bergmasco 位于阿达河（River ADDA）上游，冬季的清晨，浓雾笼罩着整个世界。我们在家里用木炭生起火炉，大家一起准备着温暖而丰盛的菜肴，以抵御寒冷和潮湿。

如果太阳出来，阳光洒进来，迷雾消散时，我们喜欢去树林里采摘栗子，这也是家乡的一项传统。我喜欢在火上烤栗子或用它们来煮汤或做甜点，多么美好的食物！

November is cold months in Italy. My village in Italy, Cisano Bergmasco, sits on the upper River Adda, and at this time of year, it is surrounded by thick fog every in morning. At home, we turn on our wood stove and prepare warm and hearty dishes to fend off the cold and damp.

If the sun burns off the fog, we love to go to the woods and harvest chestnuts, which is also a tradition in my hometown. We like to roast the chestnuts on the fire or boil them to make soups and desserts. They truly are a wonderfully versatile ingredient!

Menu

羽衣甘蓝煎蛋饼
Frittata con Cavolo Nero
Kale Frittata

栗子汤配香煎鸭肝
Zuppa di Castagne
Chestnut Soup with Foie Gras

意式藏红花烩饭
Risotto allo Zafferano
Saffron Risotto

炖牛小排、玉米糊及玛莎拉酱
Brasato di manzo e Polenta
Braised Short Ribs, Polenta and Marsala sauce

栗子奶油
Castagne
Chestnut Cream

FRITTATA CON CAVOLO NERO
KALE FRITTATA

羽衣甘蓝煎蛋饼

羽衣甘蓝 300 克	300g kale
洋蓟 200 克	200g artichokes
洋葱 50 克	50g onions
萨拉米香肠 100 克	100g salame
鸡蛋 6 个	6 eggs
帕玛森芝士 50 克	50g Parmigiano Reggiano
橄榄油 50 克	50g olive oil
黄油 80 克	80g butter
牛奶 50 毫升	50ml milk
山羊奶酪 50 克	50g goat cheese
盐适量	Salt, to taste
黑胡椒适量	Black pepper, to taste

制 作	将羽衣甘蓝洗净，在冷水中冲洗几次，切成小块。
	洋蓟清洗干净，去掉硬叶和茎，保留内芯，切成四瓣，放入柠檬水中保鲜。

制　　作　　将羽衣甘蓝洗净，在冷水中冲洗几次，切成小块。

洋蓟清洗干净，去掉硬叶和茎，保留内芯，切成四瓣，放入柠檬水中保鲜。

洋葱切碎。萨拉米香肠和山羊奶酪切成小块。

将鸡蛋放入搅拌碗中，加入帕玛森芝士、盐和黑胡椒，搅拌均匀。

在平底锅中加入 2 勺橄榄油和黄油，加入洋葱碎，翻炒大约 5 分钟直到洋葱变软，放入洋蓟和羽衣甘蓝，加入一勺水，煮 10 分钟，关火后静置冷却。

把鸡蛋和煮好的蔬菜放入烤盘，表面刷上适量黄油，撒上适量帕玛森芝士，放入烤箱 200℃烤 10 分钟左右。将食材从烤盘中取出，趁热上桌。

Prepare the dish

Wash the kale well, rinse it a couple of times in cold water and cut into julienne.

Clean the artichokes, removing the hard leaves and the stems and keeping the heart. Cut into quarters and set aside in lemon water to keep fresh.

Chop the onions. Cut the salami into small cubes and do the same with the goat cheese.

Crack the eggs into a mixing bowl and mix well with Parmigiano Reggiano, salt, and black pepper.

In a sauté pan, pour two spoons of olive oil and the butter, then add the onions, and cook until soft about 5 minutes. Add the artichoke and the kale with a spoonful of water and cook for 10 minutes then set aside to cool.

Transfer the eggs and the boiled vegetables to an oven tray brushed with some soft butter, sprinkle some Parmigiano Reggiano over the top and cook for about 10 minutes in a 200°C oven. Remove from the tray and serve hot.

ZUPPA DI CASTAGNE
CHESTNUT SOUP WITH FOIE GRAS

栗 子 汤 配 香 煎 鸭 肝

黄油 100 克	100g butter
特级初榨橄榄油 100 克	100g extra virgin olive oil
大葱 100 克，切丝	100g leeks, julienned
鸡汤 200 毫升	200ml chicken stock
烤栗子 800 克，去皮，切碎	800g roasted chestnuts, peeled and sliced
波特酒 100 克	100g Port wine
鸭肝 300 克，切块	300g foie gras, diced
盐适量	Salt, to taste
黑胡椒适量	Black pepper, to taste
罗勒油适量	Basil oil, to taste

汤的制作　在汤锅中加热黄油和橄榄油，加入切好的大葱丝。
当葱变软时，加入一勺鸡汤。随即加入栗子，用盐和黑胡椒调味。
再加入波特酒，熬煮至酒液蒸发。如果需要，可以再加些鸡汤。
将汤搅拌至润滑的乳液状后，用细筛过滤。保温放置。
在平底锅中，将切好的鸭肝煎至表面脆黄、中间温热，然后切成方块，保温放置。
把汤倒入切好的鸭肝中。用烤好的栗子和几滴罗勒油装饰。

Prepare
the soup
Heat the butter and olive oil in a soup pan and add the leeks.
When the leeks are soft, add one spoon of chicken stock. Now add the chestnuts and season with the salt and the black pepper.
Sprinkle with the Port wine and let the alcohol evaporate. If needed, add some more chicken stock.
Blend to a smooth emulsion and strain through a very fine sieve. Keep warm.
In a sauté pan, cook the slices of foie gas until crispy and brown outside and warm on the inside. Cut into cubes and keep warm.
Serve the soup and top with the cubes of foie gras. Finish with some slices of crispy chestnut and some drops of basil oil.

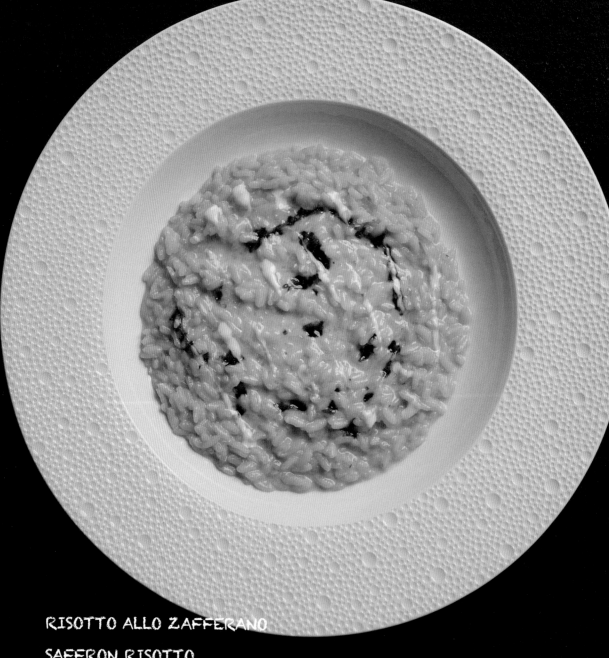

RISOTTO ALLO ZAFFERANO

SAFFRON RISOTTO

意式藏红花烩饭

鸡汤 400 毫升	400ml chicken stock
藏红花 3 克	3g saffron
小葱 50 克，切丝	50g shallots, finely diced
意大利米 250 克	250g risotto rice
白葡萄酒 80 克	80g white wine
黄油 150 克	150g butter
新鲜现磨帕玛森芝士 150 克	150g freshly grated Parmigiano Reggiano
特级初榨橄榄油 50 克	50g extra virgin olive oil
盐适量	Salt, to taste
黑胡椒适量	Black pepper, to taste

制　　作　　将鸡汤煮沸，加入藏红花。

另取锅，倒入橄榄油，油热后放入葱丝，当葱丝颜色呈金黄色时，加入意大利米和适量白葡萄酒。待酒液挥发，一勺一勺地加入鸡汤，用木勺一直搅拌直至被米吸收。大约 12 分钟后，尝一下米饭口感，如果你喜欢可以加少许黄油。

将锅离火，加入余下的黄油和帕玛森芝士，用黑胡椒、盐和橄榄油调味即可。

Prepare
the dish

Bring the chicken stock to the boil and add the saffron.

Heat the shallots in another pan with the olive oil. When the shallots are golden, add the risotto rice and sprinkle with the white wine. When the alcohol has evaporated, start to slowly add the chicken stock, ladle by ladle, stirring constantly with a wooden spoon until absorbed by the rice. After 12 minutes, check if the rice is al dente, and add a little butter if you like.

Remove from the heat and finish with the remaining butter and Parmigiano Reggiano. Season with black pepper and salt to taste and finish with olive oil.

炖牛小排、玉米糊及玛莎拉酱

红葡萄酒 500 克	500g red wine
玛莎拉葡萄酒 500 克	500g Marsala wine
特级初榨橄榄油 60 克	60g extra virgin olive oil
芹菜 200 克	200g celery
胡萝卜 200 克	200g carrots
洋葱 200 克	200g onions
牛小排 1 公斤	1kg beef short ribs
芥末适量	Mustard, to taste
盐适量	Salt, to taste
现磨黑胡椒适量	Freshly ground black pepper, to taste
小牛肉高汤 500 毫升	500ml veal stock
黄油 60 克	60g butter
玉米粉 200 克	200g polenta
水 1 升	1L water

| 牛 小 排 | 在锅中倒入红葡萄酒和玛莎拉葡萄酒，加热蒸发至剩余三分之一液体。
将芹菜、胡萝卜和洋葱切成小块，放入锅内，加入橄榄油翻炒，直至蔬菜呈漂亮的金黄色。
将牛小排用盐、黑胡椒和少许芥末调味。在平底锅内倒入橄榄油并加热，待油温非常高时，将牛小排的每一面煎熟。
在一口铸铁锅里放入炒好的蔬菜、牛小排、熬煮后的葡萄酒和小牛肉高汤，盖上锅盖，文火煮 4小时。煮好后，将牛小排从锅里捞出，放置在托盘里。
把锅里剩下的酱汁倒入搅拌机中搅匀，如果想再浓稠些，可以将黄油加热后放入一同搅拌。 |

| Prepare the beef | Combine the wine and the Marsala wine in a pot and reduce to 1/3 of the volume.
Roast the celery, carrots and onions in a pan with some olive oil, making sure that they take on a nice golden brown colour .
Now, take the beef short ribs and season well with the salt, the black pepper, and some mustard. Heat a sauté pan with some olive oil and when the oil is very hot, sear the meat on every side .
Now place the vegetables, meat, wine reduction, and veal stock in a large cast iron pot, cover the pot and cook for 4 hours over a gentle flame. When the beef is ready, remove it from the pot and place it in a tray.
Blend the sauce in a blender. If it needs reducing more, whisk in a little hot fresh butter. |

| 玉 米 糊 | 在一口汤锅中倒入 1 升水，加入盐、黑胡椒和一勺橄榄油，当水沸腾时加入玉米粉，用手持搅拌棒不停地搅拌 10 分钟。
在盘中放一勺玉米糊，放上一块牛小排并淋上酱汁。 |

| Prepare the polenta | Bring 1L water to the boil in a large pot and season with salt, black pepper and a spoon of olive oil. When the water is boiling, add the polenta and whisk constantly for 10 minutes with a whisk.
Place a generous spoon of polenta on each plate, carve a slice of beef, and cover with the sauce. |

CASTAGNE
CHESTNUT CREAM

栗 子 奶 油

海绵焦糖蛋糕 | **For the caramel sponge**
白砂糖 (A) 200 克 | 200g sugar (A)
水 80 毫升 | 80ml water
蛋黄 150 克 | 150g egg yolks
蛋糕粉 110 克 | 110g cake flour
杏仁粉 100 克 | 100g almond meal
蛋白 180 克 | 180g egg whites
白砂糖 (B) 100g | 100g sugar (B)

朗姆酒糖浆 | **For the rum syrup**
白砂糖 100 克 | 100g sugar
水 250 毫升 | 250ml water
黑朗姆酒 80 克 | 80ml dark rum

栗子奶油 | **For the chestnut cream**
栗子泥 800 克 | 800g chestnut puree
牛奶 800 毫升 | 800ml milk
白砂糖 120 克 | 120g sugar
转化糖浆 80 克 | 80g Trimoline

栗子冰激凌 | **For the chestnut gelato**
牛奶 1000 毫升 | 1000ml milk
奶油 320 克 | 320g cream
白砂糖 100 克 | 100g sugar
葡萄糖粉 100 克 | 100g dextrose powder
稳定剂 6 克 | 6g stabiliser
蛋黄 100 克 | 100g egg yolks
栗子奶油 600 克 | 600g chestnut cream
黑朗姆酒 30 毫升 | 30ml dark rum

栗子奶油泡芙 | **For the chestnut Chantilly**
奶油 300 克 | 300g cream
糖粉 30 克 | 30g icing sugar
栗子奶油 200 克 | 200g chestnut cream

海绵焦糖蛋糕	锅中加入水和白砂糖（A），熬煮成棕色的焦糖，静置备用。待焦糖放凉后，将其倒入蛋黄中并充分搅打，直到蛋黄变得轻盈蓬松。将蛋糕粉单独过筛，再与杏仁粉混合。在打发的蛋黄中，分三次加入面粉混合物，注意不要消泡。混合好后，加入融化的黄油，搅拌均匀。 用白砂糖（B）和蛋白制作蛋白酥，搅打至中等打发程度，然后加入到蛋糕面糊中翻拌。将面糊铺在烤盘上，放入烤箱180℃烘烤15分钟。烤制完成后从烤箱中取出，稍微冷却后用刀切出数个圆形海绵蛋糕。将这些蛋糕放入冰箱冷藏，直至可以盛装甜品为止。

Prepare the caramel Sponge

In a small pot, cook the water and the first portion of sugar (A) to a caramel of a medium brown colour. Set aside. Once the caramel is cooled, pour it onto the egg yolks and start to whip until they become very light and fluffy. Separately, sift the cake flour and mix it with the almond meal. Once the yolks are whipped, fold into them the flour mixture in three batches, taking care to preserve the foam. Add the melted butter and mix it through. Separately, make a meringue with the second portion of sugar (B) and the egg white, whipping to medium soft peaks. Fold it into the sponge batter. Spread the batter on a baking tray and bake it at 180°C for 15 minutes. Once baked, take it out of the oven and let it cool down a little before cutting some disks out with a cutter. Reserve the disks into the fridge until ready to plate the dessert.

朗姆酒糖浆

用小锅把白砂糖融化成糖浆。在使用前将其从火上取下，加入朗姆酒，以保持多元的风味。

Prepare the rum syrup

In a small pot, make a syrup by melting the sugar into the water. Once melted, take it off the heat. Add the rum just before using to preserve the alcohol flavour.

栗子奶油

小锅中加热牛奶，放入白砂糖和转化糖浆。待糖融化后，将锅从火上移开，慢慢将牛奶倒在栗子泥中，搅拌均匀。将奶油保存在密封盒中，放入冰箱冷藏。

Prepare the chestnut cream

In a small pot, warm the milk and add the sugar and the Trimoline. Once the sugars are melted, take the mixture off the heat and slowly pour the milk onto the chestnut puree and mix thoroughly. Keep the cream in a sealed box in the fridge.

栗 子 冰 激 凌	取一口中等大小的锅，把牛奶和奶油混合后加热。将白砂糖、葡萄糖粉和稳定剂放入碗中搅拌。把蛋黄放入温热的牛奶里，然后将干性食材放入，不断搅拌，注意不要结块。加热至82℃后关火，将栗子奶油和黑朗姆酒放入锅中，搅拌均匀，即成冰激凌底料。把冰激凌底料放入冰箱里冷藏一晚。第二天将冰激凌做好，再放入冰箱里冷藏。

Prepare the chestnut gelato	In a medium sauce pan, mix the milk with the cream and start to warm the mixture. In a bowl, mix the sugar with the dextrose and the stabiliser. Pour the egg yolks into the warm milk and then sprinkle in the dry ingredients, taking care to not form clumps. Cook the base to 82℃, then take off the heat. Add the chestnut cream and the dark rum and mix thoroughly with a hand blender to form the gelato base. Set aside for one night in the fridge. The next day, churn the gelato and then set aside in the refrigerator.

栗子奶油 泡　　芙	把所有原料混合在一起，放入搅拌机中搅拌。之后装入裱花袋中，放入冰箱里冷藏。

Prepare the chestnut Chantilly	Mix all the ingredients together and whip in a stand mixer. Transfer to a pastry bag fitted with a decorative nozzle and reserve in the fridge.

镀 层 和 装　　饰	取一片焦糖海绵蛋糕，将它浸入朗姆酒糖浆中，摆放在盘子中间。用冰激凌勺舀一个栗子冰激凌球，在蛋糕中间放一块小的栗色糖霜。将冰激凌球放在海绵蛋糕上，在其周围用栗子奶油泡芙装饰。 在另一个裱花袋中放一些栗子奶油，然后在甜点的顶部铺上栗子"面条"。在"面条"上再放一小块栗色糖霜，然后在整个甜点表面撒上糖粉。在盘子里可多摆放一些栗色糖霜。

Plating and assembly	Take one disk of the caramel sponge and soak it into the rum syrup. Place it in the middle of the plate. With an ice cream scoop, scoop a ball of chestnut gelato and place a small marron glacé in the middle. Place the gelato ball onto the sponge and decorate it with the chestnut Chantilly. Put some chestnut cream in another pastry bag and pipe chestnut "noodles" all over the top of the dessert. Place another small piece of marron glacé on the top of the "noodles" and dust the dessert with icing sugar. Decorate the plate with some more marron glacé pieces and serve.

Winter
冬

十二月

Dicembre
December

这一年即将结束，十二月最重要的节日当然是圣诞节了。在圣诞节到来之前，很多意大利人都会提前休假，大多数北方人会去滑雪，度过一个"白色周末"。还有不容错过的圣诞市集，在米兰、都灵甚至更北方的皮埃蒙特，有着全欧洲最美的圣诞市集。自家酿制的腊肠、火腿、奶酪、圣诞装饰、手工巧克力、新年礼物、家居用品等，可谓应有尽有，令人眼花缭乱，许多欧洲其他国家的市集举办者都会跑到意大利来参观学习。喝上一杯热红酒，在如画的市政厅广场边吃边逛，为家人朋友购买礼物，为圣诞节装饰居室而精挑细选，简直再开心不过了。

The most important holiday in December is, of course, Christmas. As the year draws to a close, many Italians tend to take an early holiday before Christmas. Most northerners will go skiing for a "white weekend". In big cities such as Milan, Turin, and even Piedmont, there are the most colourful Christmas markets in the world which cannot be missed. You can find everything here: homemade sausages, ham, cheese, Christmas decorations, handmade chocolates, New Year's gifts and household goods... many market organisers from other European countries even come here to visit and learn. Walking in the picturesque city square, you can enjoy delicious food, browse the stalls, shop for family and friends and pick out Christmas decorations while warming your hands with a cup of mulled wine. It is the most enjoyable time of the year.

圣诞晚宴自然是这一年中最隆重奢华的大餐了，举家一起准备最好的食物和美酒，为了让全家人开心和满足。在圣诞节当天，我们家会从下午一直吃到深夜，妈妈和祖母早早就准备好了一份长长的菜单：鹅肝、鱼子酱、白松露、牛排等，用最上等的食材来犒劳一家人这一年的辛苦，跟中国人过年吃年夜饭一样意义非凡。当然，圣诞节晚宴的重头戏当然是意大利圣诞水果面包 Panettone 了，这可是圣诞节的灵魂所在，同中国人春节吃年糕或饺子一样，寓意吉祥如意，幸福美满。

The Christmas feast is certainly the most important of the year and is celebrated with the best food and wine to make the whole family happy and satisfied. For me, on that day, I will be with my family and enjoy nice food from the afternoon to late at night. My grandmother and my mother would prepare a long menu list early. Foie gras, caviar, truffles, and steak all have their place on the table. Choice food is the best reward for a year's hard work. It's as meaningful as the family reunion dinner at Spring Festival. It certainly wouldn't be Christmas without Panettone, the fruit bread only for Christmas. Italians eat Panettone while Chinese have rice cakes and dumplings. It's a tradition that brings good luck and happiness.

Menu

腌香橙三文鱼配酸奶油及意大利鱼子酱
Salmone arancia Caviale Calvisius
Orange-Marinated Salmon with Sour Cream and Caviar

香煎鸭肝及塔加斯卡橄榄和糖渍柠檬
fegato gasso D'ANATRA e Limone candito
Roasted Foie Gras with Taggiasca Olives and Candied Lemon

手工意式细面配意大利阿尔巴白松露泡沫
Tagliolini al Tartufo bianco
Homemade Tagliolini with White Truffle from Alba

米兰风味炸小牛排及芝士汁
Vitello alla Milanese e fonduta di Formaggi
Veal Chop Milanese with Cheese Fondue

意大利圣诞面包
Panettone

SALMONE ARANCIA CAVIALE CALVISIUS
ORANGE-MARINATED SALMON WITH
SOUR CREAM AND CAVIAR

腌香橙三文鱼配酸奶油及意大利鱼子酱

橙子 5 个	5 oranges
柠檬 4 个	4 lemons
白砂糖 250 克	250g sugar
盐 200 克	200g salt
Ora 皇家三文鱼鱼柳 400 克	400g Ora King salmon fillets
鱼子酱 40 克	40g caviar
橙子酱汁 50 克（详见一月食谱）	50g orange dressing (see January recipes)
酸奶油 60 克	60g sour cream

制　作　用擦屑器将橙子和柠檬的外皮擦成丝。在碗中将橙子丝和柠檬丝混合，用盐和白砂糖调味，再挤入一个柠檬和一个橙子的汁。

将三文鱼的鱼皮去掉后放入烤盘中，把刚才调好味的柠檬橙子丝覆盖在三文鱼上，放入冰箱冷藏腌制 8 小时。

8 小时后，把柠檬橙子丝去掉，将三文鱼切成条。

用酸奶油、一勺鱼子酱和适量橙子酱汁装饰盘子，最后放上三文鱼即可。

Prepare the dish

Using a microplane, zest the oranges and lemons. In a bowl, mix the lemon and orange zest with the salt and sugar, and add the juice of 1 lemon and 1 orange.

Place the salmon on a tray, and cover all over with the aromatic salt. Marinate in the fridge for 8 hours.

After 8 hours, remove the zest from the salmon and cut into thin slices.

Place the salmon on a plate and garnish with sour cream, a spoon of caviar, and some orange dressing.

FEGATO GASSO D'ANATRA E LIMONE CANDITO
ROASTED FOIE GRAS WITH TAGGIASCA OLIVES AND
CANDIED LEMON

香煎鸭肝及塔加斯卡橄榄和糖渍柠檬

柠檬 1 个	1 lemon
塔斯马尼亚蜂蜜 50 克	50g Tasmanian honey
塔加斯卡橄榄 100 克	100g Taggiasca olives
番茄干 50 克	50g sun-dried tomatoes
法棍面包 1 个，切片	1 baguette, sliced
鸭肝 350 克	350g foie gras
橄榄油适量	Olive oil, to taste

制　　作　　将柠檬去皮，然后将柠檬果肉最大程度地切碎。用沸水将柠檬皮烫三次。
在碗中，将柠檬皮、柠檬果肉和蜂蜜混合均匀，备用。
将橄榄去核，与番茄干一同切碎，搅拌成橄榄酱（tapenade）。
将法棍面包涂抹上橄榄油，放入烤箱中烤制，直至颜色呈金黄后取出，在面包上放上满满的橄榄酱。
在热锅内煎制鸭肝，确保鸭肝表面略脆、内芯温热，然后将鸭肝放在厨房纸上。
先将法棍面包铺在盘子上，随后放上鸭肝，用蜂蜜柠檬皮点缀即可。

Prepare
the dish

Peel the skin from the lemon in strips. Cut the lemon flesh into supremes. In a pot of boiling water, blanch the lemon skin three times.
Mix the boiled lemon skin and the supremes in a bowl with the honey and set aside.
Remove the pits from the olives and dice into a tapenade together with the sun-dried tomatoes, is the tapenade.
Toast the baguette in a pan with some olive oil until it is golden. Top with a generous spoon of tapenade.
Cook the foie gras in a hot skillet, making sure it is crispy outside and warm inside, and then set aside on kitchen paper.
Place the tapenade topped baguette slices on a plate, top with the foie gras, and garnish with the honey lemon skin confit.

手工意式细面配意大利阿尔巴白松露泡沫

蔬菜高汤 100 毫升	100ml vegetable stock
黄油 100 克	100g butter
新鲜手工意式细面 350 克	350g fresh tagliolini
新鲜现磨帕玛森芝士 100 克	100g freshly grated Parmigiano Reggiano
白松露 50 克	50g white truffles
盐适量	Salt, to taste
白胡椒适量	White pepper, to taste

制　　作　　将蔬菜高汤和黄油放在平底锅中加热成酱汁。
将意式细面放在另一口锅中煮 2 分钟。
将煮好的意面放入平底锅中，加入帕玛森芝士、盐和少许白胡椒，翻拌。
将意面装盘，多浇一些酱汁，然后在上面撒上适量白松露即可。

Prepare
the dish

Heat the vegetable stock and the butter in a sauté pan.
Cook the tagliolini in another pot for 2 minutes.
Add the tagliolini to the sauté pan with the sauce and add the Parmigiano Reggiano, salt, and
some white pepper, turning and mixing.
Serve the tagliolini with lots of sauce and top with shaved white truffle.

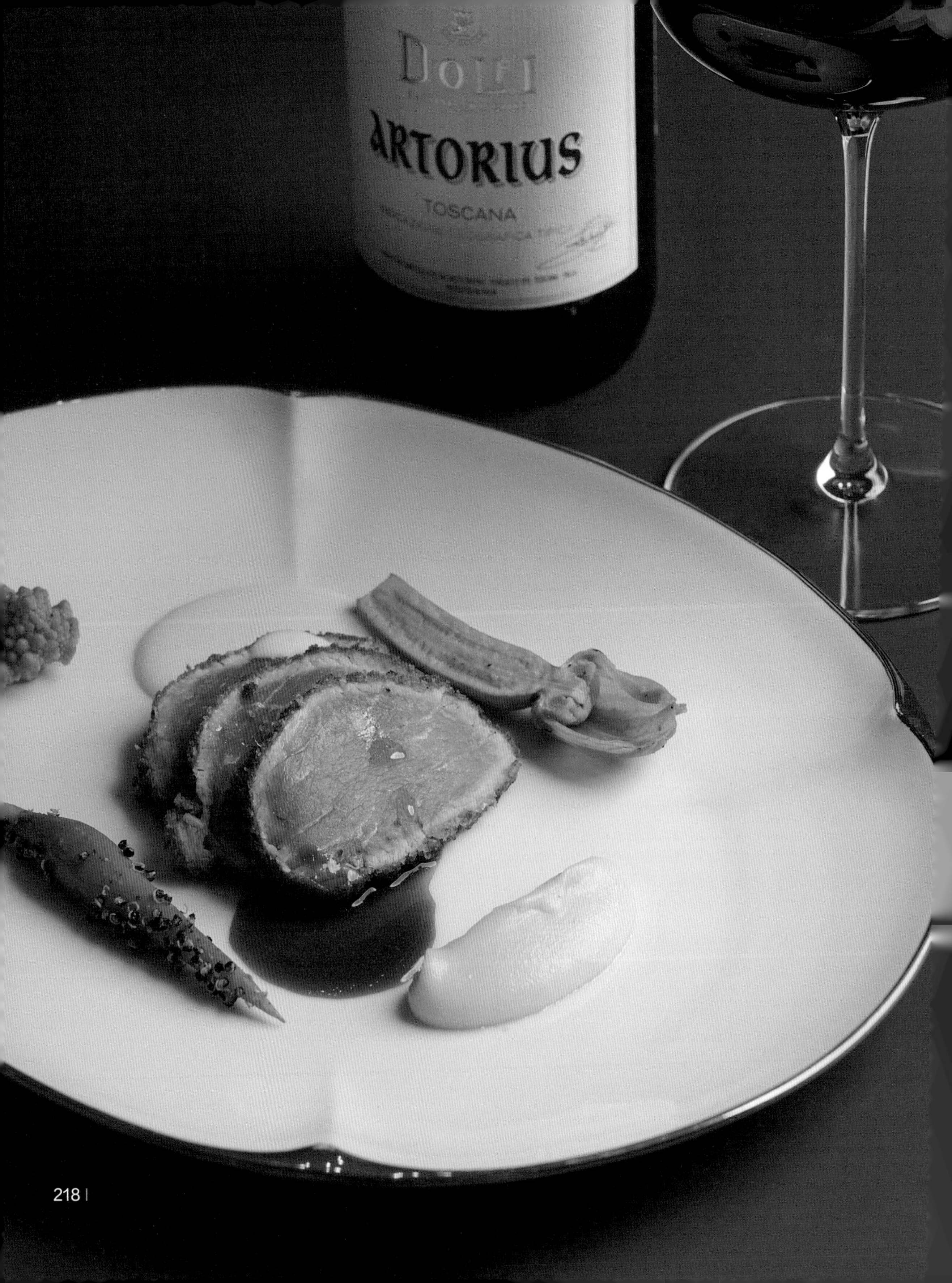
ARTORIUS
TOSCANA

VITELLO ALLA MILANESE E FONDUTA DI FORMAGGI
VEAL CHOP MILANESE WITH CHEESE FONDUE

米兰风味炸小牛排及芝士汁

芝士汁
牛奶 50 毫升
奶油 50 克
新鲜现磨蒙塔西欧乳酪 100 克
新鲜现磨帕玛森芝士 50 克

牛排
牛小排 600 克
面粉 100 克
鸡蛋 2 个
面包碎 100 克
清黄油 200 克
新鲜现磨白胡椒适量
盐适量
土豆泥适量
牛肉汁适量

For the cheese fondue
50ml milk
50g cream
100g freshly grated Montasio cheese
50g freshly grated Parmigiano Reggiano

For the veal chop Milanese
600g veal tenderloin
100g flour
2 eggs
100g breadcrumbs
200g clarified butter
Freshly ground white pepper, to taste
Salt, to taste
Mashed potato, to serve
Veal sauce, to serve

芝 士 汁　　锅中加入牛奶和奶油，加热至 65℃。加入蒙塔西欧乳酪和帕玛森芝士，均匀搅拌 5 分钟，静置保温。

For the
cheese
fondue

In a small pan, heat the milk and the cream to 65℃ . Add the Montasio cheese and Parmigiano Reggiano, and whisk well for 5 minutes, then set aside to keep warm.

制　　作　　将牛小排切成圆形块，用盐和白胡椒调味。
将面粉、鸡蛋和面包碎分别放在三个碗中。将牛小排按照面粉、鸡蛋、面包碎的顺序依次蘸过，最后压一下，确保面包碎粘在肉上。
在牛排锅中加入清黄油，加热直至冒烟，然后快速地炸牛小排。随后把油倒掉，将牛小排放在厨房纸上沥油。
将一勺土豆泥放在盘中，淋上适量牛肉汁和芝士汁，再放上牛小排即可。

Prepare
the dish

Cut the veal tenderloin into medallions and season with salt and white pepper.
Put the flour, eggs, and breadcrumbs in three separate bowls. Pass the veal medallions through the flour, then the eggs, and finally the breadcrumbs, pressing the breadcrumbs on firmly.
Heat the clarified butter in a large skillet until smoking then quickly deep-fry the medallions. Remove from the oil and drain on kitchen paper.
Place a spoon of mashed potato on each plate with some drops of veal sauce. Add the cheese fondue in the middle of the plate and top with a veal medallion.

意 大 利 圣 诞 面 包

第一步
高筋面粉 4 公斤
白砂糖 1.15 公斤
黄油 1.45 公斤
水 1500 毫升
蛋黄 1 公斤
天然酵母 1.2 公斤

For the first dough
4kg bread flour
1.15kg sugar
1.45kg butter
1500ml water
1kg egg yolks
1.2kg natural yeast

第二步
高筋面粉 1 公斤
香草荚 4 根
白砂糖 1.35 公斤
洋槐蜂蜜 250 克
蛋黄 1.3 公斤
盐 80 克
黄油 1.5 公斤
葡萄干 2 公斤
蜜饯橙皮 2 公斤
蜜饯柠檬皮 2 公斤

For the second dough
1kg bread flour
4 vanilla pods
1.35kg sugar
250g acacia honey
1.3kg egg yolks
80g salt
1.5kg butter
2kg raisins
2kg candied orange peel
2kg candied lemon peel

杏仁釉
杏仁 1 公斤
榛子 500 克
白砂糖 3 公斤
玉米淀粉 100 克
马铃薯淀粉 100 克
蛋白 1.3 公斤
珍珠糖适量

For the almond glaze
1kg almonds
500g hazelnuts
3kg sugar
100g corn starch
100g potato starch
1.3kg egg whites
Pearl sugar, to taste

第 一 步	用揉面机将高筋面粉、白砂糖、黄油、水和三分之一的蛋黄混合搅拌 15~20 分钟，直至面团成型且手感顺滑。加入天然酵母，搅拌几分钟使酵母与面团混合均匀，再将剩下的蛋黄加入面团中拌匀，直到所有的蛋黄都与面团混合在一起。制作面团的过程需要 25~30 分钟，切记不要搅拌太久。将面团从揉面机中取出，用手揉成球状，放入涂了油的容器中，放置于室温下过夜。面团在 26℃ ~28℃的理想环境中，经过大约 10~12 小时发酵，体积膨胀至原来的三倍。

Prepare the first dough	In a kneading machine, mix the bread flour, sugar, butter, water and 1/3 of the egg yolks until a silky and smooth dough is formed. This should take around 15~20 minutes. Add the natural yeast and mix for few more minutes to incorporate it to the dough. Add the remaining egg yolks and mix it into the dough. Keep mixing until all the egg yolks have been incorporated, 25~30 minutes maximum. Extract the dough from the machine, roll it into a ball and place it to proof overnight at room temperate inside a big oiled container. In an ideal environment, around 26℃ ~28℃, the dough should take around 10~12 hours to triple in volume.

第 二 步	将面团放入揉面机，放入高筋面粉和香草荚，混合搅拌 15~20 分钟，再加入白砂糖、洋槐蜂蜜和三分之一的蛋黄，搅拌至面团表面质地光滑。再次搅拌直到面团再次光滑，加入盐和三分之一的蛋黄继续搅拌。几分钟后，面团融合了所有的配料，再次变得光滑富有弹性的时候，就是加入黄油的合适时机，黄油必须保持柔软且不融化的状态，再加入剩余三分之一的蛋黄。最后加入葡萄干、蜜饯橙皮和蜜饯柠檬皮，充分搅拌，使它们均匀地分布在面团中，整个过程应不超过 35~40 分钟。 将面团从揉面机中取出，平铺在砧板上，立即将其分成小块面团，并将每块面团滚成球状，醒发 1 小时。之后再次将面团揉成球状，然后将面团放入纸模中，让它发酵到需要的程度。在 26℃ ~28℃的理想环境中，面团需要 6~8 小时才能发酵到合适的水平，发酵好的面团高度应接近纸膜边缘。

Prepare the second dough	Place the first dough into the kneading machine together with the bread flour and the seeds from the vanilla pods and start to mix. Mix until the dough returns to a smooth and silky surface and texture, around 15~20 minutes. At this stage, add the sugar and honey with 1/3 of the egg yolks. Mix the dough until smooth again, then add the salt and another 1/3 of the egg yolks and keep mixing. After a few minutes, the dough should incorporate all the ingredients and become smooth and elastic again. Now, add the butter, softened but not melted, and the remaining egg yolks. Lastly, add the raisins and candied orange and lemon peel, and mix just enough to distribute evenly through the dough. The above process should take no more than 35~40 minutes. Take the dough out of the machine and spread it on a surface. Divide it immediately into portions of your desired size and shape each portion into a ball. Leave to proof for 1 hour and then out again to a ball shape. This time place the dough into a paper mold and let it proof to the desired level. In an ideal environment, around 26℃ ~28℃, the dough should take around 6~8 hours to almost reach the border of the cup.

杏 仁 釉
和 烘 焙

使用搅拌机将杏仁、榛子、白砂糖和玉米淀粉研磨成细粉。加入蛋白，继续搅拌直到形成釉面。在将制作好的圣诞面包放入釉面之前，要先在上面加上一些杏仁釉料，并撒上一些完整的杏仁和珍珠糖。圣诞面包没有固定的烘烤时间，这取决于面团的大小，在内部温度达到 96℃时即烤制完成。烤好后，从烤箱中取出面包，用一根粗且长的铁针从面包底部穿入，以便让它保持倒置直至完全冷却。

Prepare
the
almond
glaze and
bake the
panettone

Using a blender, grind the almonds, hazelnut, sugar, and starches to a fine powder. Add the egg whites and keep blending until a glaze forms. Before putting the proofed panettone into the oven, glaze the top with some of the almond glaze, then scatter over some whole almonds and pearl sugar. There is no set baking time for the panettone as it depends on the size of the dough. Instead, it will be ready when the internal temperature reaches 96℃. After baking, take the panettone out of oven. Pierce it from the bottom with a long thick iron needle to rest it upside down to complete cool it down.

一月

Gennaio
January

一月是一年全新的开始，全世界都会为此而庆祝。新的一年，新的生活，新的目标，新的愿望……这些美好，都在一月发生。

January marks the beginning of a brand-new year and is celebrated by the whole world. A new year, a new life, new goals, new aspirations… They are all made in January.

在意大利，跟世界任何一个国家一样，呼朋引伴，举家欢聚，还有盛大晚宴，共同迎接新年。而新年晚宴的餐桌上当然要有隆重的各色海鲜以及各式各样美味的甜点。

Italy, of course, is no different. We Italians usually celebrate the beginning of the new year with a huge meal of seafood and plenty of desserts surrounded by family and friends.

一月，另外一个很重要的节日就是 1 月 6 日基督教的"主显日"（耶稣诞生日）。在这一天，我们会吃特别的糖果，看起来像炭烤的牛轧糖。

Another important date for Italians is January 6th, the Feast of Epiphany. On this day, people eat a type of candy that looks like nuggets of charcoal.

一月，对食材来说是珍贵而美好的时节，可以品尝到最美味的也是当季最后一轮的白松露。在意大利，漫山遍野白雪皑皑，人们在家里温暖的火炉旁，或欢庆、或聚会、或小酌、或饱餐，享受着食物带来的幸福以驱散严寒。

January is one of the best months for ingredients. It is also the perfect opportunity to savour the final but best white truffles. Italy is all covered with snow in January. We stay with family to celebrate, party, drink and feast by a warm fire, enjoying food to ward off the cold.

Menu

蒸新西兰鳌虾配冬季根茎时蔬沙拉
Radici invernali e Scampi al Vapore
Winter Root Salad with Steamed Scampi

新鲜洋蓟浓汤配盐鳕鱼
Carciofi e Baccala Mantecato
Artichoke Soup with Salted Cod

那不勒斯风味管面及慢炖牛肉酱
Paccheri alla Genovese
Neapolitan Paccheri with Braised Beef Ragout

澳洲牛肉、鸭肝及马德拉红酒汁
Filetto alla Rossini
Australian Beef Tenderloin, Foie Gras and Madeira Sauce

混合水果牛奶布丁
Panna Cotta Ai Frutti Rossi
Red Fruit Panna Cotta

RADICI INVERNALI E SCAMPI AL VAPORE
WINTER ROOT SALAD WITH STEAMED
SCAMPI

蒸新西兰螯虾配冬季根茎时蔬沙拉

橙子酱汁	For the orange dressing
橙子 2 个	2 oranges
鲜榨橙汁 500 毫升	500ml fresh orange juice
橄榄油 100 克	100g olive oil
蜂蜜 40 克	40g honey
白砂糖 20 克	20g sugar
黄原胶 2 克	2g xantham gum
盐适量	Salt, to taste
白胡椒适量	White pepper, to taste

沙拉	For the salad
白萝卜 100 克	100g white turnips
甜菜 100 克	100g rainbow beets
芹菜根 100 克	100g celery roots
芹菜 80 克	80g celery
茴香头 80 克	80g fennel bulbs
银鱼柳 20 克	20g anchovies
第戎芥末酱 15 克	15g Dijon mustard
白葡萄酒醋 10 克	10g white vinegar
蛋黄 1 个	1 egg yolks
橄榄油 30 克	30g olive oil
新西兰螯虾 8~10 只	8-10 pieces New Zealand scampi
海盐适量	Sea salt, to taste
白胡椒适量	White pepper, to taste

橙子酱汁	将橙子剥皮，去掉橙皮上的白色橘络，将鲜榨橙汁和橙皮一起放入锅中煮沸。待橙汁浓缩至三分之二后，与其余制作酱汁的食材一起放入搅拌机内，高速搅拌 5 分钟。将混合物过筛，得到质地柔滑的啫喱状橙子酱汁。
Prepare the orange dressing	In a pot, combine the orange juice and the skin of the orange, making sure to remove the white pith (set the orange segments aside). Bring the mixture to the boil and reduce the liquid by 1/3, then add all the remaining ingredients. Transfer to a blender and process for 5 minutes at high speed. Strain the mixture through a fine sieve to obtain a silky gel.

| 沙　拉 | 在流动水中将所有蔬菜清洗干净。用小刀将根茎类蔬菜去皮，芹菜和茴香头去筋。将蔬菜切丝，还可以将蔬菜切成不同形状，为沙拉增加乐趣。 |

在流动水中将所有蔬菜清洗干净。用小刀将根茎类蔬菜去皮，芹菜和茴香头去筋。将蔬菜切丝，还可以将蔬菜切成不同形状，为沙拉增加乐趣。

将银鱼柳切碎，过筛后碾碎成泥状。

将海盐、白胡椒、第戎芥末酱、白葡萄酒醋、银鱼柳泥、蛋黄放入一个大碗中搅拌均匀，随后缓慢倒入橄榄油制作成浓稠酱汁。

将酱汁放入盘中打底，随后放入用橙子酱汁拌好的蔬菜，再撒上少许海盐和白胡椒。

将去皮的新西兰鳌虾放入蒸笼蒸 30 秒，将其摆放在沙拉上。将剩余橄榄油刷在鳌虾上，再撒上少许海盐即可。如果你喜欢，还可以在上面点缀少许鱼子酱。

Prepare the salad

Wash all the vegetables well under running water. With a small knife, remove the skin from the roots and de-string the celery and fennel bulbs. Cut all the vegetables into julienne (cut the vegetables into different shapes for a more interesting salad).

Finely chop the anchovies with a chef's knife and pass them through a sieve to make a cream.

In a large mixing bowl, combine the salt, white pepper, Dijon mustard, white wine vinegar, anchovies, and egg yolk and whisk together. Slowly pour in the olive oil to create an emulsion.

Place some of the emulsion on the plate and toss the roots and vegetables with the orange dressing and some extra salt and white pepper.

In a steamer basket, cook the peeled langoustine for 30 seconds. Arrange on top of the salad.

Use the remaining olive oil to brush over the scampi and sprinkle with Maldon sea salt. Top the salad with a few scoops of caviar, if you like.

CARCIOFI E BACCALA MANTECATO
ARTICHOKE SOUP WITH SALTED COD

新鲜洋蓟浓汤配盐鳕鱼

盐鳕鱼 | **For the salted cod**

干盐鳕鱼 200 克　　200g dried salted cod
月桂叶 2 片　　2 bay leaves
肉豆蔻 1 颗　　1 nutmeg
牛奶 100 毫升　　100ml milk
葡萄籽油 100 克　　100g grape seed oil
白胡椒适量　　White pepper, to taste
肉豆蔻粉适量　　Nutmeg powder, to taste

洋蓟浓汤 | **For the soup**

洋蓟 500 克　　500g artichokes
橄榄油 40 克　　40g olive oil
大蒜 1 瓣　　1 clove garlic
百里香 1 撮　　Thyme, 1 pinch
黄油 50 克　　50g butter
鸡汤 100 毫升　　100ml chicken stock
甘草粉适量　　Liquorice powder, to taste
盐适量　　Salt, to taste
黑胡椒适量　　Black pepper, to taste

盐鳕鱼	将干盐鳕鱼放在冷水里浸泡一晚。

将干盐鳕鱼放在冷水里浸泡一晚。

第二天，将月桂叶、肉豆蔻、牛奶和盐鳕鱼放入锅内，加水没过食材，文火慢煮 1 小时。将煮好的鳕鱼捞出，去掉鱼皮及鱼骨。

待鳕鱼冷却后，将其放在碗中，撒上适量白胡椒和肉豆蔻粉。再慢慢加入适量葡萄籽油和煮鱼的牛奶，持续搅拌至糊状。

Prepare
the
salted
cod

The day before you want to serve the dish, soak the dried salted cod in cold water overnight.
The next day, combine the bay leaves, nutmeg, milk, and salted cod in a pot. Cover with water, bring to a simmer, and cook on a low heat for about 1 hour. Remove the fish from the aromatic milk and carefully remove the skin and bones.
When the cod is cool, transfer to a bowl and season with white pepper and nutmeg powder. Slowly add the grape seed oil and some of the aromatic milk, whisking until you reach a spoonable consistency.

洋蓟浓汤

将洋蓟洗净，把外皮剥掉，切开后去掉洋蓟芯中的杂质。

锅中倒入橄榄油，中火煸炒大蒜和百里香，再加入 25 克黄油，放入洋蓟煸炒几分钟，之后加入鸡汤。

文火煮至洋蓟变软，汤汁浓缩至原来的一半，然后使用手动搅拌器搅拌至浓稠。加入盐、黑胡椒和余下的黄油。

将汤盛在一个温热的盘子里，将鳕鱼糊做成肉丸的形状后放入汤中，撒上少许甘草粉。

Prepare
the
soup

Wash and clean the artichokes, removing the external leaves and the internal filament.
Heat the olive oil over a medium heat with the garlic clove and the thyme, then add half of the butter and the artichokes. Roast gently for a few minutes, then add the chicken stock.
Cook over a low heat until the artichokes are soft and the soup has reduced by half, then blend to a silky texture with a hand blender. Add the salt, black pepper and the remaining butter.
Serve the soup in hot bowls and finish with a quenelle of salted cod and a dusting of liquorice powder.

PACCHERI ALLA GENOVESE
NEAPOLITAN PACCHERI WITH
BRAISED BEEF RAGOUT

那不勒斯风味管面及慢炖牛肉酱

橄榄油 200 克	200g olive oil
辣椒 2 个	2 chillies
月桂叶 3 片	3 bay leaves
洋葱 250 克	250g onions, finely sliced
牛腩 200 克	200g beef brisket
第戎芥末酱 50 克	50g Dijon mustard
牛肉高汤 500 毫升	500ml beef stock
番茄酱 200 毫升	200ml tomato sauce
管状意面 400 克	400g paccheri
陈年佩科里诺羊奶干酪 100 克	100g aged Pecorino Sardo
黑胡椒适量	Black pepper, to taste
盐适量	Salt, to taste

制　　作　　在铸铁锅中，加入 2/3 的橄榄油、切碎的辣椒、月桂叶和切片的洋葱，翻炒 20 分钟，直至洋葱呈透明色。

将第戎芥末酱涂抹在牛腩上，撒上少许盐和黑胡椒，随后在平底锅内煎至表面形成一层硬壳。

将牛腩放入炒洋葱中，再倒入牛肉高汤和番茄酱，盖上锅盖，文火煨炖 4 个小时。

当牛腩煮至软嫩后，取出冷却。持续搅拌锅内的酱汁，直至质地顺滑后转为中火。

将牛腩切成 0.5 厘米厚的小块，并与酱汁混合。

另起锅，将意面放入沸水中煮熟，沥水后拌入酱汁。在意面上撒上一些牛腩块和佩科里诺羊奶干酪即可。

Prepare
the dish

In a cast iron pot, add two-thirds of the olive oil, the finely chopped chillies, the bay leaves, and the finely sliced onions. Cook the onions for 20 minutes without letting it take on too much colour.

Brush the brisket with the Dijon mustard and season with the salt and black pepper, then sear in a sauté pan to get a nice crust.

Add the piece of brisket to the onions and then add the beef stock and tomato sauce. Cover with a lid and simmer on a low heat for 4 hours.

When the meat is tender, remove it from the sauce and leave to cool. Blend the sauce until smooth and reduce the heat to medium.

Cut the meat into small cubes (0.5cm) and mix with the sauce.

Cook the paccheri in plenty of boiling water in another pot. Strain and mix with the sauce. Serve with some of the meat cubes on top and sprinkle with aged Pecorino Sardo.

澳洲牛肉、鸭肝及马德拉红酒汁

酱汁 | **For the sauce**
小葱 1 根，切碎 | 1 shallot, finely chopped
马德拉红酒 400 毫升 | 400ml Madeira wine
牛肉汁 200 毫升 | 200ml beef stock
红酒醋 1 勺 | 1 tsp red wine vinegar
甜菜汁 50 毫升 | 50ml beetroot juice
芥末籽 10 克 | 10g mustard seeds
松露水 100 毫升 | 100ml truffle water
黄油 15 克 | 15g butter

土豆泥 | **For the mashed potato**
土豆 500 克 | 500g potatoes
黄油 200 克 | 200g butter
牛奶 1 升 | 1L milk
松露碎 50 克 | 50g truffle skin
白胡椒适量 | White pepper, to taste
肉豆蔻适量 | Nutmeg, to taste
海盐适量 | Sea salt, to taste

牛肉 | **For the beef**
澳洲牛肉 4 块（200 克 / 块） | 4 pieces of Australian beef, 200g each
第戎芥末酱 1 勺 | 1 tsp Dijon mustard
大蒜 1 瓣 | 1 garlic clove
迷迭香 1 枝 | 1 spring of rosemary
鸭肝 4 片（80 克 / 片） | 4 slices of foie gras, 80g each
黑胡椒适量 | Black pepper, to taste
盐适量 | Salt, to taste
橄榄油适量 | Olive oil, to taste
黄油适量 | Butter, to taste
佩里戈尔黑松露 150 克 | 150g Périgord black truffles, to serve
应季蔬菜适量 | Seasonal vegetables, to serve

酱　　汁	将切碎的小葱放在马德拉红酒中熬煮，直至酒液浓缩至原来的三分之二时，倒出过滤。
	在一口小锅中，边搅拌边加入牛肉汁、红酒醋、甜菜汁、芥末籽和松露水，最后加入红酒浓缩汁。
	出锅前 5 分钟慢慢加入黄油，持续搅拌至质地顺滑澄亮。

Prepare the sauce

Bring the shallot to the boil with the Madeira wine. Reduce to two-thirds and strain.

In a small pan, add the beef stock, the red wine vinegar, the beetroot juice, the mustard seeds, and the truffle water. Mix with a whisk and add the Madeira reduction.

Five minutes before you are ready to serve, gradually add the butter to the sauce, whisking constantly until smooth and shiny.

土 豆 泥

将烤箱预热至 200℃。把土豆放在烤盘上，撒上适量海盐。将土豆放入烤箱烤至可以扎穿的程度。

将土豆晾至温热后去皮，用压泥器压成土豆泥。

牛奶中加入 100 克黄油、肉豆蔻、白胡椒，煮沸，再加入土豆泥，用搅拌器或木勺搅拌。当土豆泥煮至如奶油般柔滑时，借助硅胶刮刀将土豆泥过筛。

把余下的黄油和土豆泥一起搅拌，撒上松露碎。

Prepare the mashed potato

Pre-heat the oven to 200℃. Place the potatoes on a tray and sprinkle with sea salt. Place in the oven and cook until tender when pierced with a skewer.

While the potatoes are still warm, remove the skin. Pass the potatoes through a Mouli Julienne. Bring the milk to the boil with half of the butter, the nutmeg, and white pepper, then add the potators and mix well with a whisk or a wooden spoon. When the potatoes have a creamy consistency, pass them through a very fine sieve with the help of a spatula.

Finish the mashed potatoes with the remaining butter and the finely chopped truffle skin.

牛 肉

将牛肉用第戎芥末酱刷匀，撒上盐和黑胡椒。

在热的平底锅中，放入橄榄油、黄油、大蒜和迷迭香，再放入牛肉煎至你喜欢的熟度，取出静置。

把鸭肝放在不粘锅内，煎至外焦里嫩，取出放在一张厨房纸上。

盘中放入应季蔬菜、土豆泥、鸭肝、牛肉、黑松露，淋上酱汁。

Prepare the beef

Brush the beef tenderloins with the Dijon mustard and season with the salt and black pepper.

In a hot sauté pan, add the olive oil, butter, garlic, and rosemary. Cook the tenderloin to your liking then remove it from the pan and let it rest.

In a non-stick pan, roast the foie gras (it should be crisp on the outside and juicy and tender inside). Remove the foie gras from the pan and place it on a paper towel.

To serve, arrange some seasonal vegetables, a quenelle of mashed potato, the foie gras, the tenderloin, and some slices of truffle on a plate, and finish with a pool of Madeira sauce.

242

橙子焦糖
鲜榨橙汁 175 毫升
橙皮 5 克
白砂糖 250 克
水 80 毫升
盐 1 克
香草荚 1/2 根

For the orange caramel
175ml freshly-squeezed orange juice
5g orange zest
250g sugar
80ml water
1g salt
1/2 vanilla pod

香草朗姆酒意式布丁
牛奶 150 毫升
白砂糖 150 克
香草荚 1 根
吉利丁片 12 克
奶油 800 克
黑朗姆酒 30 毫升

For the vanilla and rum panna cotta
150ml milk
150g sugar
1 vanilla pod
12g gelatin
800g cream
30ml dark rum

可可碎
黄油 250 克
白砂糖 125 克
红糖 185 克
盐 1.5 克
香草荚 1/4 根
榛子粉 240 克
蛋糕粉 235 克
可可粉 45 克

For the cocoa crumble
250g butter
125g sugar
185g light brown sugar
1.5g salt
1/4 vanilla pod
240g hazelnut meal
235g cake flour
45g cocoa powder

混合浆果果汁雪葩
葡萄糖粉 150 克
白砂糖 90 克
稳定剂 4 克
水 355 毫升
草莓果泥 133 克
覆盆子果泥 133 克
新鲜蓝莓 133 克

For the mixed berry sorbet
150g dextrose powder
90g sugar
4g stabiliser
355ml water
133g strawberry puree
133g raspberry puree
133g fresh blueberries

橙子焦糖	在一口小锅里加热鲜榨橙汁和橙皮。在一口中等大小的平底锅里放入水和白砂糖，加热成浅棕色的焦糖。一旦焦糖准备好了，将刚才煮好的温热的橙汁过滤后倒入，小心蒸汽。关火，放入盐和香草荚，混合后待用。

Prepare the orange caramel	In a small pan, warm the orange juice with the zest. Place the water and sugar in a medium saucepan and cook to a light brown caramel. Once the caramel is ready, deglaze it with the warm orange juice. Be careful of the steam. Off the heat, mix in the salt and vanilla pod and then reserve the caramel until ready to assemble the dish.

意式布丁	平底锅中加热牛奶、白砂糖和香草荚。一旦糖完全融化了，牛奶也热了，放入吉利丁片，使其融化。将香草荚从混合物中取出，加入奶油，同时用手动搅拌器搅拌。最后加入黑朗姆酒，待用。

Prepare the panna cotta	In a medium saucepan, warm the milk with the sugar and the vanilla pod. Once the sugar is completely melted and the milk is warm, add the gelatin and let it melt. Remove the vanilla pod from the mixture and add the cream while mixing with a whisk. Lastly, add the dark rum. Reserve the mixture until you are ready to assemble the dish.

可 可 碎	把所有原料放在碗里，用手动搅拌器慢慢搅拌，注意搅拌均匀，不要过度搅拌。把搅拌好的面团从碗里拿出来，用烘焙切刀把它分成小块。把小块面团铺在烤盘上，放入冰箱冷藏数个小时，准备烘烤时再取出。烘烤时，将面团放入烤箱，145°C烤 20 分钟左右，具体时间取决于烤箱性能。

Prepare the cocoa crumble	Put all the ingredients in a bowl of a stand mixer and slowly mix them with the paddle attachment. Be careful to mix just until everything is incorporated and to not over mix. Take the dough out of the bowl and, with the help of a grid, break it up into small pieces. Spread the pieces out on a baking tray and reserve in the freezer for at least a couple of hours. Set the oven to 145°C, when ready, bake the crumbles for around 20 minutes, depending on your oven.

混合浆果 果汁雪葩	将葡萄糖粉、白砂糖和稳定剂放入碗中混合。 在平底锅中加热水，慢慢放入碗中的混合物，同时不断搅拌，煮至 82℃时关火。将混合物转移到罐子里，盖上保鲜膜，在冰箱里冷藏至少 12 小时。 准备做雪葩时，从冰箱里拿出罐子，加入草莓果泥、覆盆子果泥和新鲜蓝莓，立即搅拌。将所有食材搅拌均匀后放入冰箱，在 -18℃温度下冷冻。

Prepare
the
mixed
berry
sorbet

In a bowl, mix the dextrose powder, sugar, and stabiliser together.

In a medium saucepan, warm the water and start to sprinkle the dry mixture over it while stirring constantly. Cook the mixture to 82°C, then take it off the heat and transfer it to a jug. Cover with plastic film and reserve in the fridge for at least 12 hours.

When ready to make the sorbet, take the jug out of the fridge, add the strawberry and raspberry purees and the fresh blueberries and blend thoroughly with a hand blender. Churn immediately. Once ready, reserve the sorbet at -18°C.

准 备 和 摆 盘	取一个玻璃杯，在底部铺上一小层橙子焦糖，注意不要放太多，否则味道会太甜。把杯子放在冰箱里冷藏 30 分钟，让焦糖变硬，之后小心地将香草和朗姆酒撒在上面，注意不要"打碎"焦糖层，以免两种混合物混合在一起。 将杯子放回冰箱冷藏，直到奶油凝固。摆盘时，切一些新鲜的红色水果，如草莓、蓝莓、覆盆子，在表面撒上一些可可碎，然后随心情放置一些水果。在杯子中间放一勺混合浆果雪葩，你可以用新鲜的可食用鲜花或一些小蛋白酥来装饰。

Preparation
and
Plating

Take a cup or a glass and spoon a small layer of the orange caramel into the bottom. Be careful not to use too much or the dessert will be too sweet. Place the cup in the fridge for about 30 minutes to firm up the caramel. Once the caramel is ready, carefully pour the vanilla and dark rum panna cotta over it, taking care not to "break" the layer of caramel so the two mix together. Place the cup back in the fridge until the cream sets or until you are ready to serve the panna cotta. Once you are ready to serve, cut up some fresh red fruits such as strawberries, blueberries, and raspberries. Spread some cocoa crumble over the surface of the panna cotta and then place the fruit randomly. Place a scoop of mixed berry sorbet in the middle of each panna cotta. You can also decorate each cup with some fresh edible flowers, or some small meringues.

二月

一月结束，随之而来的就是冬天结束，春天开始，没错，这一切都会发生在二月。FEBRUARIUS 是二月的拉丁语名，这个词也是"净化"的意思，在意大利，二月也称为排毒月。

同时，二月还是恋人月，2月 14 日情人节是一年中最浪漫的一天，恋人们用礼物和浪漫的晚餐来庆祝爱情。对浪漫的意大利人来说，这一天当然是二月里最重要的日子。在意大利的维罗纳，这里是罗密欧和朱丽叶故事的发生地，著名的朱丽叶的阳台就在维罗纳古城内，很多游客都会在情人节这一天来到维罗纳游览，纪念和见证爱情的伟大。

As January passes, February arrives. February is the month when winter draws to a close and spring begins. "Februarius", the Latin name for February, means "Purification", so February is also the month of the detox in Italy. It is also a month for lover: the most romantic day of the year, Valentine's Day, falls on February 14. People celebrate their loved ones with gifts and romantic dinners. It is, of course, the most important day in February for romantic Italians. Verona is where the story of Romeo and Juliet took place, and the famous Juliet's balcony can still be found in the ancient city of Verona. Many tourists will come to visit here on Valentine's Day to witness and commemorate the greatness of love.

一年之际在于春，虽然二月是春天的开始，但天气仍然很寒冷，所以很多意大利人会在温暖的时节腌制一些茄子和番茄储存，到二月吃的时候，就会给厨房带来些许夏日阳光的味道。同时，二月也是许多根茎类蔬菜最甜美多汁的时节，意大利人喜欢用芹菜、红菜头、洋蓟等食材来制作健康爽口的沙拉，或者用这些蔬菜来炖汤，出锅前淋上适量橄榄油和红酒醋，鲜美极了。

Although February is the start of spring, the weather can still be cold in Italy, so many people bring a little summer sun into the kitchen by eating the eggplants and tomatoes that they preserved during the warmer months.

February is also the month when root vegetables such as celery root, beets, and Jerusalem artichokes are at their best. Italians tend to make healthy salads and soups with them, or simply boil them and drizzle them with olive oil and wine vinegar before eating. Vegetables served this way taste deliciously fresh.

Menu

澳洲小牛肉薄片配金枪鱼及水瓜柳酱
Vitello Tonnato
Veal Tenderloin with Tuna and Caper Sauce

土豆大葱汤配慢炖猪腹肉
Patate, Porri e Pancetta
Leek and Potato Soup with Braised Pork Belly

自制意式手工细面配冬季黑松露
Tagliolini al Tartufo nero
Traditional Tagliolini with Black Winter Truffle

意式牛肚配番茄酱
Trippa alla Parmigiana
Beef Tripe in Tomato Sauce

榛子奶油
Nocciola
Hazelnut Crémeux

VITELLO TONNATO
VEAL TENDERLOIN WITH TUNA
AND CAPER SAUCE

澳洲小牛肉薄片配金枪鱼及水瓜柳酱

牛肉

小牛肉 800 克
橄榄油 50 克
银鱼柳 30 克
大蒜 3 瓣
迷迭香 1 枝
白葡萄酒 50 毫升
盐适量
白胡椒适量

For the veal

800g veal tenderloin
50g olive oil
30g anchovies
3 garlic cloves
1 spring of rosemary
50ml white wine
Salt, to taste
White pepper, to taste

酱汁

小牛肉碎 400 克
金枪鱼罐头 300 克
银鱼柳 30 克
黄油 150 克
白葡萄酒 30 毫升
水瓜柳 100 克
熟蛋黄 4 个
柠檬汁 1 勺

For the sauce

400g veal tenderloin trimmings
300g canned tuna
30g anchovies
150g butter
30ml white wine
100g capers
4 boiled egg yolks
1 tsp lemon juice

牛　　肉	将小牛肉切成匀称规整的形状（切下来的碎肉留着备用）。用厨房用线把肉捆成圆柱状。
	把橄榄油、银鱼柳、大蒜和迷迭香放在铸铁锅内加热，随后放入牛肉煎至表面变色。倒入白葡萄酒，加热至酒液蒸发，随后将牛肉取出静置。
	煎好的牛肉应该呈外焦里嫩的状态，将牛肉放在架子上晾凉。

Prepare the veal	Cut the veal tenderloin into regular pieces (set aside the trimmings to use later). Use kitchen string to tie the veal into a nice cylinder.
	Heat the oil, anchovies, garlic, and rosemary in a cast iron pan and brown the veal tenderloin well on each side. Add the white wine and let it evaporate, then remove the veal tenderloin from the heat.
	The veal tenderloin should be pink in the middle with a nice golden-brown crust outside. Let the veal tenderloin rest in a cool place on a rack.

酱　　汁	把之前切下的牛肉碎切成 0.5 厘米厚的肉丁。
	铸铁锅中放入 100 克金枪鱼，再放入银鱼柳和黄油。黄油开始融化时，放入肉丁，用硅胶铲翻炒牛肉防止粘锅。随后加入白葡萄酒，加热至酒液蒸发，关火静置。
	将牛肉、鱼肉酱汁、余下的金枪鱼、水瓜柳、熟蛋黄和柠檬汁共同放入食品料理机中，搅打 10 分钟。将混合物用滤网过滤后放入冰箱冷藏。
	解开捆牛肉的绳子，将牛肉切成非常薄的片。为每一片牛肉淋上酱汁，也可以用芹菜、水瓜柳和鱼子酱来装饰。

Prepare the sauce	Cut the veal trimmings into 0.5cm cubes.
	In a cast iron pan, add 100g of tuna, the anchovies, and all the butter. When the butter melts, add the veal trimmings and scrape the bottom of the pan with a spatula. Add the white wine and let it evaporate, then remove the pan from the heat and set aside to cool.
	In a food processor, combine the meat and the caramelised juices, the remaining tuna, the capers, the boiled egg yolks, and the lemon juice. Process the mix for 10 minutes and then strain through a chinois. Taste the seasoning and keep refrigerated.
	Remove the string from the veal and slice very thinly. Serve with a generous spoon of sauce on each slice. Garnish with celery and capers, and a spoon of caviar, if you like.

PATATE, PORRI E PANCETTA
LEEK AND POTATO SOUP WITH
BRAISED PORK BELLY

土豆大葱汤配慢炖猪腹肉

泡沫
黄油 50 克
大葱 200 克
土豆 150 克，切片
奶油 200 毫升
鸡汤 250 毫升
盐适量
白胡椒适量

For the foam
50g butter
200g leeks
150g potatoes, sliced
200ml cream
250ml chicken stock
Salt, to taste
White pepper, to taste

猪腹肉
猪腹肉 500 克
百里香适量
白葡萄酒 100 毫升
胡萝卜 100 克，切丁
芹菜 100 克，切丁
洋葱 100 克，切丁
大蒜 4 瓣
口蘑 200 克，切丁
冻干的牛肝菌 125 克
鸡汤 200 毫升
盐适量
黑胡椒适量
培根碎适量

For the pork belly
500g pork belly
Thyme, to taste
100ml white wine
100g carrots, finely diced
100g celery, finely diced
100g onions, finely diced
4 garlic cloves
200g button mushrooms, finely diced
125g freeze-dried porcini
200ml chicken stock
Salt, to taste
Black pepper, to taste
Pancetta, to serve

泡　　沫　　将大葱的葱白和葱绿分开，葱白切成细丝。
用文火将黄油加热至融化，放入所有葱，缓慢加热，再放入土豆片，倒入奶油和鸡汤。
用文火煮至少 1 个小时，加入适量盐和白胡椒，搅拌至口感顺滑。将汤过滤，放置在虹吸瓶中保温。

Prepare the foam　　Separate the white and green parts of the leeks and shred the white parts.
Heat a pan over a low heat and melt the butter. Add the leeks and cook gently, then add the potatoes, the cream, and the chicken stock.
Cook over a low heat for at least 1 hour, then season with the salt and the white pepper to taste and blend to a smooth soup. Pass through a fine sieve and then put in a syphon bottle and keep warm.

| 猪 腹 肉 | 猪腹肉洗净去皮后，放入锅中，加入百里香快速翻炒。倒入白葡萄酒，待酒液蒸发后，将猪腹肉静置。
| | 另起锅，把所有蔬菜炒至略焦，放入牛肝菌和猪腹肉，再倒入鸡汤没过所有食材，加盖，文火炖3 个小时。
| | 待肉质软嫩时，将猪腹肉从汤汁中取出，放置在架子上晾凉。将蔬菜捞出控水，用盐和黑胡椒调味。
| | 用虹吸壶做出一些泡沫放在盘子上，放上一块猪腹肉和煨炖好的蔬菜，再用剩余泡沫将其覆盖。可以用一些蔬菜叶和培根碎来装饰。

| Prepare the pork belly | Clean the pork belly and score the skin with a sharp knife, then season with thyme and quickly pan fry. Sprinkle the pork belly with the white wine and once the wine has evaporated, put the pork to one side to rest.
| | Sauté the diced vegetables until golden brown. Add the porcini and the pork belly and cover with the chicken stock, ensuring the stock covers all the ingredients. Cover with a lid and simmer gently for 3 hours.
| | When the meat is tender, take it out of the liquid and leave it on a rack to cool. Reduce the vegetable liquid and season with the salt and the black pepper.
| | Charge the syphon and put some foam on the plate. Add one cube of pork belly and some of the braised vegetables, then cover with more foam. Garnish with some leaves and some crispy pancetta.

TAGLIOLINI AL TARTUFO NERO
TRADITIONAL TAGLIOLINI WITH BLACK WINTER TRUFFLE
自制意式手工细面配冬季黑松露

松露酱	For the truffle emulsion	意式细面	For the tagliolini
黄油 50 克	50g butter	鸡汤 100 毫升	100ml chicken stock
胡萝卜 100 克，切丁	100g carrots, finely diced	黄油 150 克	150g butter
芹菜 100 克，切丁	100g celery, finely diced	自制意式手工细面 350 克	350g homemade tagliolini
洋葱 100 克，切丁	100g onions, finely diced	帕玛森芝士 150 克	150g Parmigiano Reggiano
松露 60 克，切碎	60g truffle skin, finely diced	松露酱 200 克	200g truffle emulsion
红葡萄酒 50 克	50g red wine	冬季黑松露 40~50g	40-50g black winter truffles
		盐适量	Salt, to taste
		白胡椒适量	White pepper, to taste

松露酱　锅中放入 25 克黄油，文火翻炒切好的蔬菜。
加入红葡萄酒，待酒液蒸发，加入松露碎和剩余的 25 克黄油。将所有材料放入搅拌机中搅拌至顺滑。用保鲜膜盖好，保温静置。

Prepare the emulsion
Put half of the butter in a pot and gently cook the chopped vegetables over a low flame.
Add the wine and let it reduce, then add the truffle skin and the remaining butter. Transfer to a blender and blend until smooth. Keep warm, cover with clingfilm and set aside.

意式细面　把鸡汤和 75 克黄油、盐、白胡椒放入平底锅中，煮沸。
另取锅，沸水煮意面 2 分钟。
当意面煮至软硬适中时，把意面捞入鸡汤锅中，再加入帕玛森芝士和余下的 75 克黄油，混合均匀。
用勺子把松露酱盛放在温热的盘子上，再放上意面，随后刨一些冬季黑松露点缀即可。

Prepare the tagliolini
Put the chicken stock and half of the butter, salt, and white pepper in a large sauté pan and bring to the boil.
Cook the tagliolini in plenty of boiling water for about 2 minutes.
When just al dente, strain the pasta and mix it with the butter and chicken stock, then add the Parmigiano Reggiano and the remaining butter.
Spoon some of the truffle emulsion onto a warm plate, add the pasta and generously shave over the black winter truffles.

TRIPPA ALLA PARMIGIANA
BEEF TRIPE IN TOMATO SAUCE

意式牛肚配番茄酱

橄榄油 50 克	50g olive oil
洋葱 120 克，切丝	120g onions, julienned
牛肚 1.5 公斤	1.5kg beef tripe
番茄酱 220 克	220g tomato sauce
辣椒 1 个	1 chilli
牛肉高汤 250 毫升	250ml beef stock
帕玛森芝士 150 克	150g Parmigiano Reggiano
盐适量	Salt, to taste
黑胡椒适量	Black pepper, to taste

制 作

在铸铁锅中倒入适量橄榄油，放入洋葱丝，文火煸炒 10 分钟，加一大勺牛肉高汤。

把牛肚切成宽 1 厘米、长 6 厘米的长条，和洋葱丝一起在锅中煸炒 5 分钟。加入番茄酱和辣椒（可以根据口味喜好添加）。

加入适量牛肉高汤，文火熬煮 2~3 个小时。如果水分变少，可以再加一些牛肉高汤。

当牛肚炖好时，加入帕玛森芝士、盐和黑胡椒，再炖 5 分钟即可。

Prepare
the dish

Heat the olive oil in a cast iron pot then add the onions and cook over a low heat for at least 10 minutes, adding a spoonful of beef stock if the onions are browning too much.

Cut the tripe into strips about 1cm wide and 6cm long. Add to the pot with the onions and cook for 5 minutes. Add the tomato sauce and the chilli (depending on how spicy you like it).

Add some of the beef stock and cook gently for 2~3 hours. Add some more stock if the tripe gets dry.

When the tripe is tender, add the parmigiano Reggiano, salt, and black pepper, cook for another 5 minutes, then serve.

NOCCIOLA
HAZELNUT CRÉMEUX

榛子奶油

榛子奶油
吉利丁片 4 克
奶油 (A)100 克
60% 榛子酱 500 克
100% 榛子酱 200 克
奶油 (B)400 克

For the hazelnut crémeux
4g gelatin sheets
100g cream (A)
500g 60% hazelnut praline
200g 100% hazelnut paste
400g cream (B)

巧克力挞面团
黄油 (融化的) 180 克
80% 黑巧克力 145 克
白砂糖 120 克
温水 100 毫升
蛋糕粉 460 克

For the chocolate tart dough
180g butter (melted)
145g 80% dark chocolate
120g sugar
100ml warm water
460g cake flour

巧克力挞填充
55% 黑巧克力 136 克
奶油 100 克
100% 榛子酱 100 克
牛奶 100 克
鸡蛋 65 克

For the chocolate tart filling
136g 55% dark chocolate
100g cream
100g 100% hazelnut paste
100g milk
65g whole eggs

榛子碎
35% 酒心巧克力 120 克
60% 榛子酱 230 克
酥脆薄片 200 克
盐 2 克

For the hazelnut crumble
120g 35% azelia chocolate
230g 60% hazelnut praline
200g royaltine feuilletines
2g salt

榛子冰激凌
水 300 毫升
牛奶 244 毫升
白砂糖 60 克
稳定剂 6 克
转化糖 211 克
蛋黄 20 克
100% 榛子酱 100 克

For the hazelnut gelato
300ml water
244ml milk
60g sugar
6g stabiliser
211g inverted sugar
20g egg yolks
100g 100% hazelnut paste

热巧克力慕斯
蛋黄 60 克
白砂糖 60 克
奶油 150 克
68% 黑巧克力 290 克
蛋白 200 克

For the hot chocolate mousse
60g egg yolks
60g sugar
150g cream
290g 68% dark chocolate
200g egg whites

榛子奶油　将吉利丁片放入冰水中浸泡。
在平底锅中将奶油（A）加热，放入吉利丁片并融化。将两种榛子酱放入碗中混合均匀，将温热的奶油倒在上面，随后搅拌，充分乳化后倒入奶油（B），继续搅拌至完全乳化。将混合物放入裱花袋中，放入冰箱冷藏。

Prepare the hazelnut crémeux

Hydrate the gelatin sheets in an ice bath.
In a sauté pan, warm up the first portion of cream (A) and melt the gelatin. Put the hazelnut praline and paste in a tall jug and pour the warm cream over while mixing with a hand blender to emulsify thoroughly. Once it is absorbed, start to pour in the second part of the cream (b). Keep emulsifying until all of the cream has been absorbed. Transfer the mixture to a piping bag and reserve in the fridge.

巧克力 挞面团	将黄油和黑巧克力放入碗中，用微波炉加热至 50℃，充分混合后加入白砂糖。一旦糖和混合物混合均匀后，慢慢倒入温水，使其混合。将混合物倒入搅拌机中，加入蛋糕粉，搅拌均匀，注意不要过度搅拌。将面团用保鲜膜包起来，放在冰箱里冷藏。

Prepare the chocolate tart dough

Put the butter and the dark chocolate in a plastic bowl and warm to 50 °C in a microwave. Mix well and add the sugar. Once the sugar is combined into the mixture, start to pour the hot water over it and mix well to incorporate. Transfer the mixture into the bowl of a stand mixer and add the cake flour and mix until just combined, being careful to not over mix. Wrap the dough in plastic film and keep it in the fridge until ready to assemble the tart.

巧克力挞填充

在碗中放入黑巧克力、奶油、榛子酱和牛奶。将碗放入微波炉加热至巧克力融化，将所有材料混合均匀。随后加入鸡蛋，充分搅拌至完全乳化。将混合物放入裱花袋中，放在冰箱里冷藏。

Prepare the chocolate tart filling

Put the dark chocolate, cream, hazelnut paste and milk in a bowl. Put the bowl in the microwave and warm until the chocolate is melted and everything is blended together. Transfer the mixture to a jug and add the egg, then emulsify well with a hand blender. Transfer the mixture to a pastry bag and reserve in the fridge until ready to fill the tart.

榛子碎

在碗中混合酒心巧克力和榛子仁酱，然后倒入保鲜盒中，加入盐和酥脆薄片搅拌，注意不要把薄片压得太碎，混合后放入冰箱冷藏备用。

Prepare the hazelnut crumble

In a plastic bowl, mix the chocolate with the hazelnut praline, then transfer it into a sealed container. Add salt and the Royaltine feuilletines and mix without crushing them too much. Keep it in the fridge.

榛　子 冰　激　凌	平底锅里放入水和牛奶，加热。将白砂糖和稳定剂放入碗中混合，慢慢倒入牛奶锅中。随后加入蛋黄和转化糖，继续搅拌，待混合物加热至 82℃，关火。将混合物倒入罐子里，趁着还热的时候加入榛子酱，充分搅拌均匀后盖上保鲜膜，放入冰箱冷藏至少 12 小时。准备制作冰激凌时再从冰箱中取出，立即用搅拌机搅拌，在 -18℃环境下保存。

Prepare the hazelnut Gelato	Put the water and milk in a sauté pan. In a bowl, mix the sugar and stabiliser together. Start to heat the water and the milk, pouring the dry mixture in slowly. Then, add the egg yolks and the inverted sugar. Keep cooking while stirring constantly and bring the mixture to 82°C. Transfer to a jug, and while still warm, add the hazelnut paste. Mix with a hand blender, and once everything is combined, cover with plastic wrap and reserve in the fridge for at least 12 hours. When ready to make the gelato, take the jug out of the fridge, homogenise everything with the help of a hand blender and churn immediately. Once ready, keep the gelato at -18°C.

热巧克力 慕　　斯	在碗中混合蛋黄和白砂糖。锅中加热奶油，煮沸后慢慢倒入蛋黄混合物，同时用手动搅拌器搅拌。把混合物放回小锅里，加热至 82℃，注意不要把混合物煮过头，否则蛋黄会凝结，奶油状态会变差。 把黑巧克力放在罐子里，待奶油混合物加热好后，倒在巧克力上。在温度下降前，把所有食材搅拌均匀后加入蛋白，继续搅拌。用过滤器将混合物倒入虹吸壶中，使用三级力度彻底摇晃它。 如果打算马上使用它，可以把虹吸壶放在 50℃的温水中，切勿放进冰箱。如果暂时不用，则需要放在冰箱里冷藏，但记得在使用慕斯之前把虹吸壶加热到合适的温度。

Prepare the hot chocolate mousse	In a small bowl, mix the egg yolks and the sugar. Separately, in a small pot, boil the cream. Once the cream is boiling, slowly pour over the yolk mixture while stirring with a whisk. Put the mixture back in the small pot and bring it to 82°C. Be very careful to not overcook the mixture, or the yolks will curdle and the cream will not be smooth. Put the chocolate in a jug and, once the cream is ready, pour it over the chocolate. While still warm, melt the chocolate and combine everything together. Add the egg white and blend it in the mixture. Pour the mixture into a siphon using a strainer, close it and charge the gun with 3 chargers, shaking it thoroughly. If you plan to use the mousse straight away, keep the siphon out of the fridge in a warm water bath at 50°C. Otherwise, put the mousse in the fridge to use it later, but remember to warm the siphon up to the right temperature before using.

<table>
<tr><td>准 备 和
摆　　盘</td><td>将巧克力挞面团擀至约 2.5 毫米厚，制作成挞皮，放入预热好的烤箱，160℃烤 17~20 分钟，具体时间可根据你的烤箱性能调整。
烘焙完成，开始组装甜点。关闭烤箱，在烤箱内操作，将巧克力挞馅填满挞皮，挞馅的质地要硬且呈现奶油状。
选一个盘子，在盘子中间舀入适量榛子奶油，形成一条长线。拿起榛子酥，用小刀把它弄碎，做成小块，将它摆放在盘子左边。将组装好的挞切成四等份，放在盘子中心。在盘子右边舀一小勺榛子冰激凌，最后在底部滴上一小滴热巧克力慕斯，立即享用。</td></tr>
<tr><td>Preparation
and plating</td><td>Roll the chocolate tart dough out to a thickness of 2.5mm, line a tart case, and bake it in a ventilated oven at 160°C for 17~20 minutes, depending on your oven.
Once the tart is baked, you can keep it aside, but if you are ready to assemble the dessert, fill it with the chocolate filling, then switch the oven off and place the tart inside. The filling should be firm but still creamy.
Take a plate and pipe a small amount of the hazelnut crémeux and with the help of a spoon, spread it into a long line across the middle of the plate. Break the hazelnut crumble into small chunks with a knife. Place it on the left side of the centre of the plate. Cut the tart into quarters and place one in the top part from the centre of the plate. On the right, put a small quenelle of hazelnut gelato and finish the dish by piping a small drop of hot chocolate mousse on the bottom. Serve immediately.</td></tr>
</table>

老马的感谢

首先，我要对为这本书付出努力的团队致以发自肺腑的感 谢，是你们让这本书的构想得以实现！谢谢！

感谢贝加莫省奇萨诺贝尔加马斯科和泰拉莫省巴夏诺友人的大力支持。

大董先生，您不仅是中国餐饮界的传奇，更是享誉全球的 明星主厨，您能亲自为这本书撰写序言，是我莫大的荣幸，也让我多年来的梦想成真！

侯德成老师，您不仅是我的挚友，更是我在中国的亲人和导师，感谢您一直以来对我的帮助以及给予我的在烹饪上的建议和指导。

感谢 Umberto Bombana 先生这么多年来对我的引领。

Luciano Tona 先生，是你在我最需要帮助的时候给我机会。

感谢 Opera BOMBANA 餐厅的邢瑞、Tim、Vivi、Andrea、Vinz 以及全体团队成员，感谢上海 8 1/2 餐厅总经理 Gianluca，主厨 Riccardo La Perna（现于澳门 8 1/2 餐厅），以及 Raffo、Gabry、Marco、Trina、Manuel 对我一如既往的支持。

感谢 Mark P 先生、Mark Z 先生、Denis、Michael、Olivier、Frank，还有美丽的 Jessica、Marvin、Mattia、Jay、Hubert 等太古酒店集团的同事们。

张际星，没有她的策划、热爱和信念，我不可能完成这本书。

感谢姚晶、蔡乐，以及出色的美味关系团队的所有付出，感谢 Robynne 把我的蹩脚"意大利英语"转化为优美的英语。感谢冬妹和小五两位摄影师为本书拍摄的所有美食大片，是你们让我的菜品有了灵魂。

Fillippo Mazzanti，谢谢你分享在书中的所有漂亮的甜品，以及日复一日将热爱、激情和知识技术都融入工作的匠人精神。你是真正的甜品艺术家，也是我最好的朋友。

感谢本书策划编辑白兰女士对我的信任。

Fiona，我最好的另一半，如果没有她的支持和陪伴，我 根本无法在北京生活和应对身为外国人的一切挑战，她是我生命中最重要的人，她是最好的。

感谢我的妈妈 Rosanna Gandolfi，感谢她教导我如何生活得快乐和有尊严。

感谢我的姐姐 Paola 和 Sara，她们一直激励着我去追求更好的生活。

感谢赵总、Zephyr、Amy、Davide，以及宏珏集团给我这个机会，能够成为 GIADA Garden 迦达花园餐厅的行政主厨，这是我生命中最美好的缘分之一。

感谢香格里拉酒店集团的 Stephan Kapek 先生给予我的一切支持。

感谢 Owen、Paolino、XiaoXiao、Tong、Yang、Luis、Felix1、Felix2、Alice、Bruce、Zoe、David、Adam、Jhon、Rikka、Eva，以及 GIADA Garden 迦达花园餐厅整个厨房团队的努力和付出。

感谢 Ashlee、Anthony、Nick、Alex、Frank、Lily、HaoRan、Ella、Lucy、Catherine、Curie、Owen、Kevin、Echo、Adrian、Lisa、Eddie，以及 GIADA Garden 迦达花园餐厅全部服务及运营团队的善良和贴心。

最后，感谢这十五年来我生活在中国遇见的所有好朋友的支持和鼓励，是你们让我有动力和勇气，继续前行。

I want to say thank you from the bottom of my heart to the people who made this book possible! Grazie!

I would like to thanks the friends from Cisano Bergamasco province of Bergamo and the friends from Basciano Province of Teramo.

Chef Da Dong is a legend both in China and around the world. It is a great honour and a dream come true for me to have his words in this book.

Mr. Hou Decheng has been more than just a great friend to me. He is also my teacher, and I would like to thank him for always being there to help me or offer opinions on my cooking.

Thank you to Chef Umberto Bombana for his guidance over the years.

Thank you to Chef Luciano Tona, who gave me a chance when I needed it the most. Thank you!

Thank you to Rain, Tim, Vivi, Andrea, Vinz and all the Opera BOMBANA team, and to Riccardo La Perna(now in Macao 8 1/2), Raffo, Gabry, Marco, Trina, Manuel, and Gianluca from 8 1/2 Otto e Mezzo Bombana Shanghai. Thank you for supporting me as always.

Thank you to Mark P, Mark Z, Denis, Michael, Olivier, Frank, Jessica (la piú bella di tutte), Marvin, Mattia, Jay, Hubert, and the Swire Hotel team.

Thank you to Jessie, because without her coordination, passion, and commitment, I couldn't have completed this book.

Thank you to Mio, Caicai and the incredible StarGourmet Communications team, and to

Robynne for transforming my words into proper English. Thank you also to Dong Mei and Mr. Wu. Their wonderful images are the quintessence of this book, bringing life to my dishes.

Thank you to Filippo Mazzanti for the beautiful desserts he prepares with passion and skill every day. He is a true master of the pastry kitchen and a great friend.

Thank you to my editor Ms. Bai Lan for her trust in me.

Love and thanks to Fiona. Without the support of my better half, I wouldn't have been able to cope with my hectic day-to-day life in Beijing. She is simply the best.

Thank you to my mother, Rosanna Gandolfi, who taught me how to live happily and with dignity.

Thank you to my sisters Paola and Sara for always motivating me to do better.

Thank you to Mr. Zhao, Zephyr, Amy, Davide, and all the beautiful people of Redstone Group for giving me the opportunity to be the executive chef of GIADA GARDEN. To me, it is much more than a restaurant. It is a wonderful life experience.

Thank you to Mr. Stephan Kapek from Shangri-La Group for his constant support.

Thank you to Owen, Paolino, XiaoXiao, Tong, Yang, Luis, Felix 1 and Felix 2, Alice, Bruce, Zoey, David, Adam, Jhon, Rikka, Eva, and all the GIADA Garden kitchen team for being so passionate and inspiring.

Thank you to Ashlee, Anthony, Nick, Alex, Frank, Lily, HaoRan, Ella, Lucy, Catherine, Curie, Owen, Kevin, Echo, Adrian, Lisa, Eddie and the front of the house team at GIADA Garden for your kindness and patience.

Finally, a big, big thank you to all the friends and colleagues who have supported me on my China adventure over the last 15 years.

版权贸易合同登记号 图字：01-2022-4897

图书在版编目（CIP）数据

意大利的四季滋味 / (意) 马里诺·安托尼奥(Marino D'Antonio) 著；张际星译. 一北京：电子工业出版社，2022.8
ISBN 978-7-121-43977-3

Ⅰ.①意… Ⅱ.①马… ②张… Ⅲ.①菜谱－意大利 Ⅳ.①TS972.185.46

中国版本图书馆CIP数据核字(2022)第129641号

责任编辑：张瑞喜
文字编辑：白 兰
印　　刷：中国电影出版社印刷厂
装　　订：中国电影出版社印刷厂
出版发行：电子工业出版社
　　　　　北京市海淀区万寿路 173 信箱　　邮编：100036
开　　本：710×1000　1/16　印张：17　字数：392 千字
版　　次：2022 年 8 月第 1 版
印　　次：2022 年 8 月第 1 次印刷
定　　价：98.00 元

凡所购买电子工业出版社图书有缺损问题，请向购买书店调换。若书店售缺，请与本社发行部联系，联系及邮购电话：（010）88254888，88258888。

质量投诉请发邮件至 zlts@phei.com.cn，盗版侵权举报请发邮件至 dbqq@phei.com.cn。

本书咨询联系方式：bailan@phei.com.cn，（010）68250802。